Electron Lenses for Super-Colliders

Particle Acceleration and Detection

springer.com

The series *Particle Acceleration and Detection* is devoted to monograph texts dealing with all aspects of particle acceleration and detection research and advanced teaching. The scope also includes topics such as beam physics and instrumentation as well as applications. Presentations should strongly emphasise the underlying physical and engineering sciences. Of particular interest are

- contributions which relate fundamental research to new applications beyond the immeadiate realm of the original field of research

- contributions which connect fundamental research in the aforementionned fields to fundamental research in related physical or engineering sciences

- concise accounts of newly emerging important topics that are embedded in a broader framework in order to provide quick but readable access of very new material to a larger audience.

The books forming this collection will be of importance for graduate students and active researchers alike.

Series Editors:

Alexander Chao
SLAC
2575 Sand Hill Road
Menlo Park, CA 94025
USA

Christian W. Fabjan
CERN
PPE Division
1211 Genève 23
Switzerland

Frank Zimmermann
CERN
SL-Division
AP Group
1211 Genève 23
Switzerland

More information about this series at http://www.springer.com/series/5267

Vladimir D. Shiltsev

Electron Lenses
for Super-Colliders

 Springer

Vladimir D. Shiltsev
Accelerator Physics Center
Fermi National Accelerator Laboratory
Batavia, IL, USA

ISSN 1611-1052
Particle Acceleration and Detection
ISBN 978-1-4939-3315-0   ISBN 978-1-4939-3317-4   (eBook)
DOI 10.1007/978-1-4939-3317-4

Library of Congress Control Number: 2015953777

Springer New York Heidelberg Dordrecht London

Printed on acid-free paper

Springer Science+Business Media LLC New York is part of Springer Science+Business Media
(www.springer.com)

This book is dedicated to the memories of Gennady F. Kuznetsov (1938–2011) and David Wildman (1950–2014)— our late colleagues, who made possible the first Tevatron electron lenses and, thus, this book.

Preface

The intent of this book is to give a comprehensive overview of the electron lenses—a novel instrument for high-energy particle accelerators, particularly for the energy-frontier superconducting hadron colliders (which are often called "supercolliders"). Three such colliders were built—the Tevatron at Fermilab, the Relativistic Heavy Ion Collider at BNL, and the Large Hadron Collider at CERN—and each of these machines represents an epoch in particle physics research. While construction of the 87-km-long Superconducting Super Collider in Texas was terminated in 1993, concepts of even larger proton-proton colliders are being actively developed now in China, Europe, and in the USA. The supercolliders are arguably the most complex research instruments ever built, and they are widely recognized for many technological breakthroughs and numerous physics discoveries. Their complexity and typically very high cost, call for superb performance and high luminosity of these machines as much is desired in the return on the investment. As the result, many advances in accelerator physics and technology have been implemented at the supercolliders, including the electron lenses—the subject of this book. The electron lenses have been proposed and employed for compensation of the beam-beam effects and for collimation of the high-energy high-intensity beams. Also, the use of the electron lenses for compensation of the space-charge effects and other accelerator applications is being actively pursued both theoretically and experimentally.

In this book, I coherently describe the technology and the physics of the electron lenses for high-energy hadron colliders and present theoretical and experimental works to date in uniform fashion. Throughout the text, I use the same symbol definitions and provide references which are readily available for the reader. For example, all the references to the proceedings of the International, European, and IEEE article accelerator conferences (PACs) can be found at JACOW website http://accelconf.web.cern.ch/accelconf/. All cited Fermilab technical publications are available at *inSPIRE* http://inspirehep.net/. Many articles from leading accelerator science and technology open-access journals, such as *Physical Review Special Topics: Accelerators and Beams* (*PRST-AB*) http://journals.aps.org/prstab/

and *Journal of Instrumentation* (*JINST*) http://iopscience.iop.org/1748-0221/, are cited throughout the book.

In Chap. 1, I outline the basics of the colliding beams technique, a brief history of the hadron superconducting supercolliders, main beam dynamics challenges they confront, and an overview of the electron lens method and its applications to address the issues on the way to high performance of such accelerators. The technology of the electron lenses—from subsystems to beam diagnostics and integration—is presented in Chap. 2, with the Tevatron and RHIC electron lenses used for illustration. Other chapters are devoted to specific applications of the electron lenses, such as for compensation of the head-on and long-range beam-beam effects (Chap. 3), for beam halo collimation (Chap. 4), and for the space-charge compensation and other proposed uses (Chap. 5).

About a hundred scientists and engineers were involved in the development of the electron lenses, their construction and installation, experimental beam studies, operations, analysis, and upgrades. These efforts span almost two decades and many leading accelerator centers including Fermilab, BNL, ORNL, ERN, JINR (Dubna), Budker Institute of Nuclear Physics, and IHEP (Protvino), and to them many thanks are due. Two of our late colleagues deserve special appreciation: Gennady F. Kuznetsov (1938–2011) and David Wildman (1950–2014), both of Fermilab, who led the design and development of the first Tevatron electron lenses being in charge of the electromechanical system integration and high-power pulsed HV systems, respectively. This book is dedicated to their memory.

Finally, I would especially like to thank Dr. Frank Zimmermann of CERN for his encouragement to write and publish this book in the *Particle Acceleration and Detection* series, Margaret Bruce for careful reading of the manuscript, and Springer's editorial team for their patience and valuable help.

Batavia, IL, USA   Vladimir D. Shiltsev
July 2015

Contents

Symbols

Symbol(s)	Meaning	Units
L	Luminosity per IP	$\text{cm}^{-2}\,\text{s}^{-1}$
$I_L = \int L\, dt$	Integrated luminosity	pb^{-1}/wk, fb^{-1}
$E_{p,a}$	Proton(antiproton) beam energy	GeV
$C, R = C/2\pi$	Ring circumference, radius	m
$c = 2.9979 \times 10^8$ m/s, v $\beta = v/c;\ \gamma_{p,a,e} = (1 - \beta^2_{p,a,e})^{-1/2}$	Speed of light, velocity, relativistic factors (protons, antiprotons, electrons)	
$f_0 = C/v, f_{RF}, h = f_{RF}/f_0$	Revolution frequency, RF frequency, harmonics number	MHz
x, y, z, s	Horizontal, vertical and longitudinal displacements, longitudinal coordinate	
$Q_{x,y,s}$	Horizontal, vertical, synchrotron tune	
N_B	Number of bunches	
T_B	Bunch spacing	ns
$N_{p,a,e}$	Protons (antiprotons, electrons)/bunch	10^9
$\varepsilon_{(p,a)\,(x,y,L)}$	RMS normalized emittance (proton, antiproton) (horizontal, vertical, longitudinal)	$\pi\,\mu\text{m}$, eV s
$\sigma_{(p,a,e)(x,y,z)}, \sigma_{E,\delta}$	RMS beam size (proton, antiproton, electron) (horizontal, vertical, longitudinal), energy spread, relative energy spread	μm, m
$\sigma_{E,\delta}$	Energy spread, relative energy spread	
$\beta_{x,y}, \alpha_{x,y}, D_{x,y}, \beta^*_{x,y}$	Beta- and alpha-beam optics functions (horizontal, vertical), dispersion, beta-function at IP	m, cm
$x,y,z = (2J_{x,y,z}\beta_{x,y,z})^{1/2}\cos(\psi_{x,y,z})$	Coordinates, action variables, and phases	
V_{RF}	RF voltage amplitude	MV
$Z_{//}, Z_{\perp}, Z_0$	Longitudinal, transverse impedance, free space impedance	Ohm, Ohm/m
		$377\,\Omega$

(continued)

(continued)

Symbol(s)	Meaning	Units
$F(\sigma_z, \beta^*, \ldots)$	Hourglass factor	
$e = 2.71828\ldots$		
$e = 1.602 \times 10^{-19}\,\mathrm{C}$	Electron charge	
$m_p = 938.27\,\mathrm{MeV/c^2}$	Proton mass	
$m_e = 511\,\mathrm{keV/c^2}$	Electron mass	
$r_p = e^2/m_p c^2 = 1.535 \times 10^{-18}\,\mathrm{m}$	Proton classical radius	
$r_e = e^2/m_e c^2 = 2.818 \times 10^{-15}\,\mathrm{m}$	Electron classical radius	

Parameters of Superconducting Hadron Colliders

(*TeV*-Tevatron, *SSC*-Superconducting Super Collider, *RHIC*-Relativistic Heavy Ion Collider, *LHC*-Large Hadron Collider, *FCC*-Future Circular Collider, *p*-protons, *a*-antiprotons, *i*-ions)

		TeV	SSC	RHIC	LHC	FCC	
Particles		*p-a*	*p-p*	*p-p, i-i,* *p-i*	*p-p*	*p-p*	
Circumference	C	6.28	87.12	3.83	26.7	~100	km
Peak energy (protons)	E	0.98	20	0.25	7	~50	TeV
Peak luminosity	L	4.3	10	2.5 (*p-p*)	100	~500	10^{32} cm^{-2} s^{-1}
Max. dipole field	B	4.4	6.6	3.5	8.3	~16	T
Injection energy	E_{inj}	0.15	1.0	0.028	0.45	~3.3	TeV
Energy ramp time		84	1000	220	1200	–	s
RF harmonic number	h	1113	104,544	2520	35,640	–	
Transition gamma	γ_t	18.6	105	22.9	55.7	–	
Maximum RF voltage	V_{RF}	1.4	20	4.0	8	16	MV
$\beta_{\max}$ in cells (h/v)		100	305	57	180	–	m
β^* at collision points		0.28	0.5	1.0	0.55	1.1	m
Maximum dispersion	D_x	8	1.8	1.9	2.0	–	m
Tune (approx.)	$Q_{x,y}$	20.59	123.28	28.68	64.31	~120	
Bend magnet length		6.1	12.7	9.45	14.3	–	m
Half-cell length		29.7	90	15	53.5	~100	m
Bend magnets/cell		8	12	2	6	–	
Bend magnets total		774	10,288	396	1232	–	
Phase advance/cell		68	90	90	90	–	deg
Cell type		FODO	FODO	FODO	FODO	FODO	

Abbreviations

AA	Antiproton Accumulator at Fermilab
BBC	Beam-beam compensation
BINP	Budker Institute of Nuclear Physics, Russia
BNL	Brookhaven National Laboratory, USA
BPM	Beam position mionitor
CERN	European Organization for Nuclear Research, Switzerland
DESY	Deutsches Elektronen-Synchrotron Laboratory, Germany
FCC	Future Circular Collider study hosted at CERN
Fermilab	Fermi National Accelerator Laboratory, USA
FNAL	Fermi National Accelerator Laboratory, USA
HEBC	Hollow electron beam collimator
HEP	Hign energy physics
HERA	Hadron-Elektron-Ring-Anlage at DESY
IHEP	Institute for High Energy Physics, Russia
IP	Interaction point
ISR	Intersecting storage rings at CERN
JINR	Joint Institute for Nuclear Research, Russia
LHC	Large Hadron Collider at CERN
Linac	Linear accelerator
MI	Main injector synchrotron at FNAL
Quad	Quadrupole magnet
RHIC	Relativistic Heavy Ion Collider at BNL
RR	Recycler Ring at FNAL
Sp(p)S	Super Proton (antiproton) Synchrotron at CERN
SSC	Superconducting Super Collider, USA
TEL	Tevatron electron lens
TMCI	Transverse mode-coupling instability
Tevatron	TeV Proton-Antiproton Collider at Fermilab

Chapter 1
Introduction

1.1 Colliding Beams

Particle accelerators have been widely used for physics research since the early twentieth century and have greatly progressed both scientifically and technologically since then. To gain insight into the physics of elementary particles, one accelerates them to a very high kinetic energy, let them impact on other particles and detect products of the reactions that transform the particles into other particles. It is estimated that in the post-1938 era, accelerator science has influenced almost 1/3 of physicists and physics studies and on average contributed to physics Nobel Prize-winning research every 2.9 years [1]. Colliding beam facilities which produce high-energy collisions (interactions) between particles of approximately oppositely directed beams did pave the way for progress since the 1960s.

The center of mass (c.o.m.) energy E_{cm} for a head-on collision of two particles with masses m_1, m_2 and energies E_1 and E_2 is

$$E_{cm} = \left[2E_1E_2 + \left(m_1^2 + m_2^2\right)c^4 + 2\sqrt{E_1^2 - m_1^2c^4}\sqrt{E_2^2 - m_2^2c^4} \right]^{1/2}. \quad (1.1)$$

For many decades, the only arrangement for accelerator experiments was a fixed target setup in which a beam of particles accelerated with a particle accelerator hit a stationary target set into the path of the beam. In this case, as follows from (1.1), for high energy accelerators $E \gg mc^2$, the CM energy is $E_{cm} \approx (2Emc^2)^{1/2}$. For example, 1 TeV = 1000 GeV protons hitting stationary protons can produce reactions with about 43 GeV c.o.m. energy. A more effective colliding beam set-up in which two beams of particles are accelerated and directed against each other, has much higher center of mass energy of $E_{cm} \approx 2(E_1E_2)^{1/2}$. In the case of two equal mass of particles (e.g., protons and protons, or protons and antiprotons) colliding with the same energy E of 1000 GeV, one gets $E_{cm} = 2E$ or 2000 GeV. Such an obvious

© Springer Science+Business Media New York 2016
V.D. Shiltsev, *Electron Lenses for Super-Colliders*, Particle Acceleration and Detection, DOI 10.1007/978-1-4939-3317-4_1

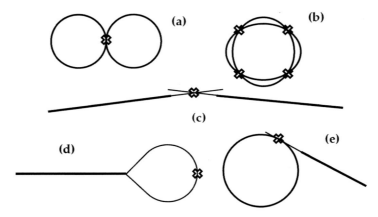

Fig. 1.1 Schematics of particle collider types (from [3])

advantage secured the place of the colliders as the frontier high energy physics machines starting in the 1960s [2, 3].

Almost three dozen colliders reached the operational stage between the late 50s and now. Schematic drawings of several collider types are shown in Fig. 1.1. In storage ring configurations—Fig. 1.1a, b—particles of each beam circulate and repeatedly collide. This can be done in a single ring if the beams consist of the same energy antiparticles. In linear colliders, first proposed in [4], the beams are accelerated in linear accelerators (linacs) and transported to a collision point either in the simple two linac configurations depicted in Fig. 1.1c, or with use of the same linac and two arcs as in Fig. 1.1d. Another possible linac-ring configuration is shown in Fig. 1.1e.

The first lepton ($e-e-$ or $e-e+$) colliders were built in the early 1960s almost simultaneously at three laboratories: AdA collider at the Frascati laboratory near Rome in Italy, VEP-1 collider in the Novosibirsk Institute of Nuclear Physics (USSR) and the Princeton-Stanford Colliding Beam Experiment at Stanford (USA). In all three colliders their center of mass energies were 1 GeV or less. Construction of the first hadron (proton-proton) collider, the Intersecting Storage Rings, began at CERN (Switzerland) in 1966, and in 1971, this collider was operational and eventually reached $E_{cm} = 63$ GeV. In the case of electrons and positrons, the synchrotron radiation results in fast damping of betatron and synchrotron oscillations and creates an effective way to accumulate large currents. The synchrotron radiation used to be negligibly weaker in handron colliders and, therefore, construction of proton-antiproton colliders required a different damping mechanism. The invention of the stochastic cooling technique in 1969 led to the construction of the first proton-antiproton collider $Sp\overline{p}S$ at CERN in 1982. The Tevatron proton-antiproton collider [5] was the world's highest energy collider for almost 25 years since its operation began in December of 1985 until it was overtaken by the LHC in 2009 [6].

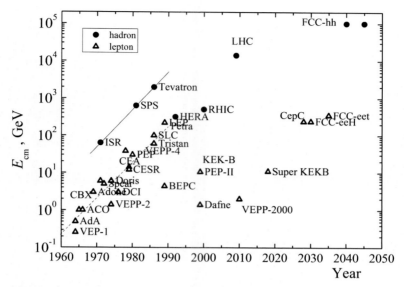

Fig. 1.2 Energy reach of particle colliders (adapted from [3])

The energy of colliders has been increasing over the years as demonstrated in Fig. 1.2. There, the triangles represent maximum CM energy and the start of operation for lepton colliders (usually, $e + e-$) and full circles are for hadron (protons, antiprotons, ions, proton-electron) colliders. One can see that until the early 1990s, the c.o.m. energy on average grew by a factor of 10 every decade and, notably, the hadron colliders were 10–20 times more powerful than the lepton ones. Since then, following the demands of high energy physics, the paths of the colliders diverged. The record high energy Large Hadron Colider (LHC) was built at CERN, while new $e + e-$ colliders called "particle factories" were focused on detailed exploration of phenomena at much lower energies.

Hadron circular colliders employing superconducting (SC) high-field magnets—often called *supercolliders*—so far have dominated the energy frontier. The reason is that the superconducting magnet technology offered the highest energy reach at an affordable cost. In the supercolliders, as in any circular accelerator, the maximum momentum and energy of ultra-relativistic particle is determined by the radius of the ring R and average magnetic field B of bending magnets:

$$pc = eB \cdot R \quad \text{or} \quad E[GeV] = 0.3 \cdot B[T] \cdot R[m]. \tag{1.2}$$

Therefore, the maximum energy reach depends on by practical considerations: e.g., maximum field of normal conducting magnets of about 2 T at some moment was not adequate for the energy demands because of required longer accelerator tunnels and increasing magnet power consumption. The development of superconducting magnets [7]—see Fig. 1.3—which employ high electric current carrying NbTi

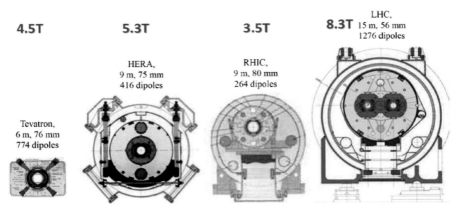

Fig. 1.3 Superconducting dipole magnets for high energy hadron colliders: Tevatron (NbTi, warm-iron, small He plant, 4.5 K), HERA (NbTi, Al collar, cold iron), RHIC (simple and economical design) and LHC (2 K super fluid He, double bore) (from [3])

wires cooled by liquid Helium below 5 K, opened the way to higher fields and record high energy hadron colliders. The latest of them, 14 TeV c.o.m. energy LHC at CERN uses 8.3 T double bore magnets in 26.7 km circumference tunnel.

To remain superconducting, such magnets need to operate within very strict limits on the power deposited into the low-temperature components (vacuum pipes, cold iron, SC cable, etc.)—typically on the order of 1 W/m or less, and that makes them of no practical use in high energy lepton accelerators, as relativistic electrons and positrons quickly lose energy due to the synchrotron radiation. The energy loss per turn is:

$$\delta E = \frac{4\pi}{3} \frac{e^2 \gamma^4}{R} = 88.5 \left[\frac{keV}{turn}\right] \cdot \frac{E^4[GeV]}{R[m]} \qquad (1.3)$$

The total power radiated into the walls becomes prohibitive at high energies and reaches, e.g., 22 MW or about 800 W/m even in the largest radius $e + e-$ collider LEP (in the same tunnel, which is now occupied by LHC), at the beam energy of 105 GeV and relatively low average beam current of 4 mA. Besides the need to replenish the electron beam power loss by accelerating RF cavities, the synchrotron radiation leads to significant heating and outgassing of the beam vacuum pipe. That makes difficult the attainment of sufficiently long lifetimes of continuously circulating beams which requires the residual gas pressures of 1–10 nTorr or better. This technological challenge has been successfully resolved in modern colliding "factories" operating with lower energy multi-Ampere beams. Radiation of protons (ions) is smaller by a significant factor $(\gamma_p/\gamma_e)^4 = (m_e/m_p)^4$—see (1.3)—but still can be of a certain concern at very high energy, high current SC accelerators, e.g., LHC.

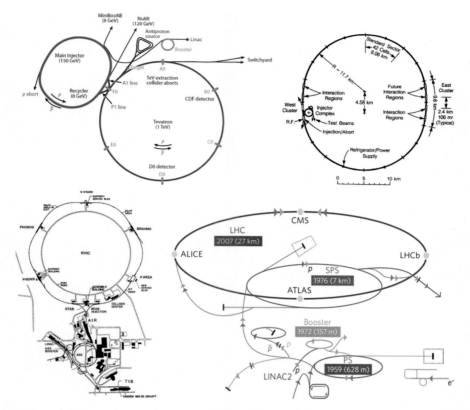

Fig. 1.4 Layouts of the superconducting hadron supercolliders (*clockwise*, from *top left*): the Tevatron (from [5]), SSC (from [9]), LHC (from [10]), and RHIC (from [8])

Though only three hadron supercolliders reached operational stage; the Tevatron, the Relativistic Heavy Ion Collider (RHIC) [8], and LHC (see Fig. 1.4) there were two notable examples of very large supercollider projects in the past, namely, the Superconducting SuperCollider in Texas, USA [9] and the UNK collider in Protvino, Russia [11]. Due to various reasons, mostly attributed to their scale, cost and complexity, both were terminated in 1990s [12]. Even larger *p-p* supercolliders were proposed—the Very Large Hadron Collider (VLHC) in the US in early 2000s [13], and, more recently, the Future Circular Collider (FCC) at CERN and the Super Proton-Proton Collider (SppC) in China [14]—see Table 1.1.

Despite the fact that these accelerators were and are the largest and the most expensive instruments for high-energy physics research, they are quite cost-effective. For example, a comparative analysis [15] of publicly available costs for 17 large accelerators of the past, present and those currently in the planning stage, has shown that the "total project cost (TPC)" (sometimes cited as "the US accounting") of a collider can be broken up into three major parts corresponding to "civil construction", "accelerator components", "site power infrastructure" and the three

Table 1.1 Hadron superconducting supercolliders and their construction costs (for the SSC, RHIC and LHC—see [15] and references therein, for the Tevatron and UNK—[16, 17], correspondingly)

	Energy (TeV, c.o.m.)	Circumference (km)	Cost (year)	Comments
Tevatron	1.96	6.3	0.45B$ (1986)	Operational, 1985–2011
SSC	40	87.1	11.8B$ (1993)	Terminated in 1993
UNK	6	20.8	~2B$ (1991)	Project stopped in 1998
RHIC	0.5	3.8	0.66B$ (2000)	Operational, 2000–present
LHC	14	26.7	6.5BCHF (2009)	Operational, 2008–present
FCC	100	100	?	~2035 ?

corresponding cost components can be parameterized by just three parameters—the total length of the accelerator tunnels L, the c.o.m. or beam energy E, and the total required site power P. It was found that over almost three orders of magnitude of L, 4.5 orders of magnitude of E and more than two orders of magnitude of P the "$\alpha\beta\gamma$-cost model" works with ~30 % accuracy:

$$Total \ Project \ Cost \approx \alpha \times Length^{1/2} + \beta \times Energy^{1/2} + \gamma \times Power^{1/2} \qquad (1.4)$$

where coefficients $\alpha = 2\text{B\$}/(10 \text{ km})^{1/2}$, $\gamma = 2\text{B\$}/(100 \text{ MW})^{1/2}$, and accelerator technology dependent coefficient β is equal to 10 B\$/TeV$^{1/2}$ for superconducting RF accelerators, 8 B\$/TeV$^{1/2}$ for normal-conducting (warm) RF, 1 B\$/TeV$^{1/2}$ for normal-conducting magnets and 2 B\$/TeV$^{1/2}$ for SC magnets (all numbers in 2014 US dollars)—see Fig. 1.5. Given that SC magnets provide a factor of ~5 higher magnetic fields compared to warm, normal-conducting magnets (and correspondingly shorter tunnels) and that they allow significant reduction of the facility electric power e, it is not surprising to see that the supercolliders offer the highest energy reach within limited financial resources. Still, very high total costs of the supercolliders usually call for maximum machine performance (luminosity, see next section), various measures to reduce the cost (e.g., extensive R&D on the cost-effective magnets [18] and tunneling, re-use of the existing infrastructure and accelerators as injectors, etc.) and, often, the expansion of the physics program beyond p-p collisions (e.g., RHIC and LHC collide ions as well as protons).

1.2 Luminosity and Beam Dynamics Issues in Hadron Supercolliders

The exploration of rare particle physics phenomena requires appropriately high energy but also a sufficiently large number of detectable reactions. In a collider, the number of the events of interest per second depends on the cross section σ_{event} of the reaction under study is given by:

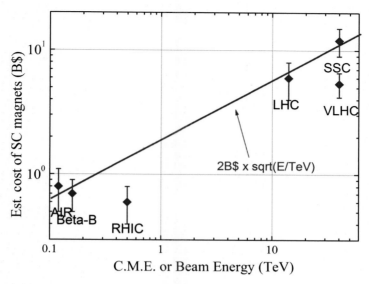

Fig. 1.5 Estimated cost of the superconducting magnets and associated elements vs. collider c.o.m. energy or single beam energy. Stage I of the VLHC assumed low-field 2 T superferric magnets (from [15])

$$dN_{event}/dt = L \cdot \sigma_{event}, \tag{1.5}$$

where L is the collider luminosity. The luminosity depends on the machine and beam parameters and for Gaussian beams is equal to:

$$L = \gamma f_0 \frac{N_B N_{p,1} N_{p.2}}{4\pi \beta^* \varepsilon} F(\sigma_s, \beta^*, \theta), \tag{1.6}$$

where $N_{p(1,2)}$ is the number of particles per bunch in each of the two colliding beams, N_B is the number of bunches per beam, ε is the average rms normalized emittances of two round beams $(\varepsilon_{p1} + \varepsilon_{p2})/2$, $F < 1$ is the geometric luminosity reduction factor. γ is the relativistic factor, and f_0 is the revolution frequency. Usually, for proton-proton colliders bunch intensities and emittances are about the same $N_{p1} = N_{p2} = N_p$ and $\varepsilon_{p1} = \varepsilon_{p2} = \varepsilon_p$, while that is not true for beams of different species, e.g., in the Tevatron proton-antiproton collider $N_p \gg N_a$ and $\varepsilon_p \gg \varepsilon_a$. The geometric luminosity reduction factor F takes into account the "hour-glass effect" (variation of the transverse beam size along the luminous region) which depends on the ratio of the rms bunch length σ_s and beta-function at the collision point β^*, and also accounts for reduction due to crossing angle θ between two beam orbits at the interaction point (IP). For example, for round Gaussian beams with $\sigma_s \ll \beta^*$ and equal transverse rms beam sizes σ^* at the IP for both beams:

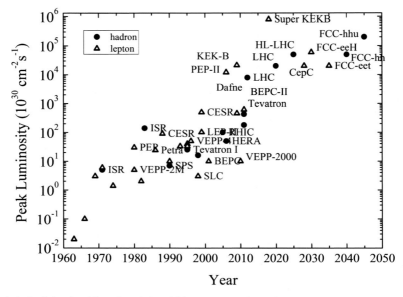

Fig. 1.6 Peak luminosities of particle colliders (adapted from [3])

$$F = \frac{1}{\sqrt{1 + \left(\frac{\theta\sigma_s}{2\sigma^*}\right)}}.$$ (1.7)

Therefore, to achieve high luminosity, one has to maximize population of bunches, minimize their emittances and to collide them as frequently as possible at the locations where the focusing beam optics provides the smallest values of the amplitude functions $\beta^*_{x,y}$. Increasing the beam energy and thus, factor γ in (1.6), is, generally speaking, of help, too, though the task of the higher energy beam focusing requires stronger and more challenging magnets in the machine's interaction regions (IRs).

Figure 1.6 demonstrates the impressive progress of luminosities of colliding beam facilities. Again, the triangles are for lepton colliders and full circles are for hadron colliders. One can see that over the past five decades, the performance of the colliders has improved by more than six orders of magnitude and reached record high values of over 10^{34} cm^{-2}s^{-1}. At such luminosity, one can expect to produce, e.g., 100 events over 1 year of operation (about 10^7 s) if the reaction cross section is minute 1 femtobarn (fb) = 10^{-39} cm^2. Table 1.2 summarizes the peak luminosity values and key beam parameters for proton-(anti)proton supercolliders.

The total number of particle reactions is, of course, proportional to the luminosity integral $I = \int L dt$ and in the end, this is the most critical parameter for collider experiments. The integral depends on the total accelerator running time (typically—many years, decades for the supercolliders), the machine availability (ratio of the luminosity operation time to calendar time), and evolution of the luminosity in

Table 1.2 Luminosity and beam parameters of proton-proton/proton-antiproton superconducting supercolliders

	E_{cm} (TeV)	Peak L, (10^{34} cm^{-2}s^{-1})	$N_{p(a)}$ (10^{11})	N_B	$\varepsilon_{p(a)}$ (μm)	β^* (cm)	ξ	W (MJ)
Tevatron	1.96	0.043	2.9/0.8	36	3/1.5	28	0.025	1.7
SSC	40	0.1	0.075	17,240	1	50	0.004	418
UNK	6	0.1	3	348	7.5	150	0.005	50
RHIC	0.5	0.025	1.9	111	3.1	65	0.018	0.8
LHC	14	1.0	1.15	2808	3.7	55	0.01	360
FCC	100	5.0	1	10,600	2.2	110	0.01	8400

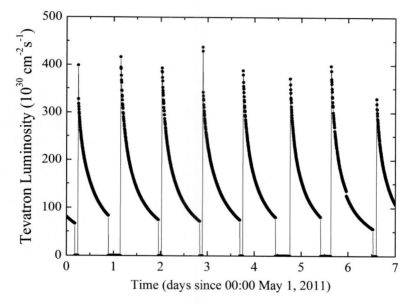

Fig. 1.7 One week luminosity record of the Tevatron proton-antiproton collider (May 1–May 8, 2011), indicating individual HEP stores and characteristic luminosity decay in each store (see text)

individual luminosity runs (sometimes also referred to as "HEP stores"). Usually, the luminosity of a hadron collider decays with some characteristic time τ_L that may vary from a few hours to dozen(s) of hours depending on the machine—see Fig. 1.7. Correspondingly, the integrated luminosity for the HEP stores longer than τ_L scales approximately as a product of the initial luminosity and the lifetime $I = \int L dt \approx L_0 \tau_L$.

Optimal performance of the supercolliders requires maximum integrated luminosity and minimizing the experimental background in the detectors (e.g., events created by the beams due to reasons different from the main collisions, such as scattering on the residual gas molecules, fast diffusion, etc.). Numerous issues which need to be addressed by accelerator designers and operators of the supercolliders are comprehensively presented and analyzed elsewhere—see, e.g., books [19–21] and reviews [5, 22]—and below we briefly overview only major luminosity limitations in superconducting hadron colliders. They can be schematically divided in two categories—effects due to collisions (beam-beam effects, events pile-up, luminosity lifetime, etc.) and single-beam effects (single-particle stability and dynamic aperture (DA), impedance and collective effects, electron cloud effects, machine stability and noises leading to emittance growth, collimation and machine protection, particle production and cooling, space-charge effects, polarization, etc.).

1.2.1   Beam-Beam and Other Effects Due to Collisions

In a hadron collider, the colliding beams must have small transverse dimensions to reach high luminosity—see (1.6). This leads to a strong electromagnetic force exerted by each beam on the particles of the other—see Fig. 1.8. The beam-beam interactions usually result in strong enhancement of particle losses and emittance growth and in reduction of the luminosity lifetime and luminosity integral (sometimes—in significant and fast beam intensity losses causing quench of superconducting magnets). They are known to be one of the most severe limitations on the collider performance and are comprehensively studied and widely discussed - see, e.g., proceedings of corresponding workshops and several review articles [23–29]. Table 1.2 indicates that maximum *beam-beam parameter* ξ—the figure of merit of beam-beam interaction that approximately equals to the shift of the particles' betatron frequency (tune) due to the collisions—which could be operationally achieved in the hadron supercolliders colliders is limited to about:

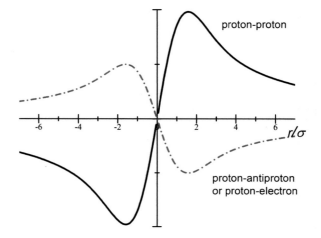

Fig. 1.8 Schematic dependence of the beam-beam force on proton vs. the proton coordinate at the IP: it is de-focusing for proton-proton interaction (*solid line*) and focusing for interaction of proton with opposite charge beam, e.g., antiprotons, electrons—see *dashed line* (scaled by factor of 2 for illustration purpose)

$$\xi = N_{IP}\zeta_{IP} = N_{IP}\frac{N_p r_p}{4\pi\varepsilon} \leq 0.025, \tag{1.8}$$

where $r_p = e^2/mc^2 = 1.53 \cdot 10^{-18}$ m denotes the classical proton radius, N_p and ε are the opposite bunch intensity and emittance, correspondingly, and N_{IP} is the total number of head-on IPs in the collider.

Contrary to $e-e+$ colliders which usually enjoy fast synchrotron radiation damping of the betatron oscillations and where attempts to overcome the maximum beam-beam parameter of $\xi \sim 0.05$–0.1 lead to threshold-like performance degradation, the beam-beam limit in the hadron colliders is usually "soft" and indicates an approximate boundary between "optimal" and "barely operationally tolerable" conditions like particle losses, detector backgrounds, emittance growth and luminosity degradation rates, etc. Still, given that hadron machines are quite sensitive to minor betatron tune variations $dQ_{x,y} \sim 0.001$ off the optimal machine working point (Q_x, Q_y), it is remarkable how carefully these colliders' parameters are set to accommodate large tune spread of the order of ξ. For example, in the Tevatron collider, particles' vertical and horizontal tunes were between the fifth and seventh order resonances (between $Q_{x,y} = 3/5 = 0.6$ and $Q_{x,y} = 4/7 = 0.571$, the integer part of the tune $[Q_{x,y}] = 20$) and the beam-beam tune spread fully covered the available tune area shown in Fig. 1.9.

An additional complication comes from operation with many bunches—dozens to thousands as indicated in Table 1.2—which is needed to address the so-called "event pile-up" phenomenon, characteristic for hadron colliders. Due to large values of the total inelastic cross sections of proton-proton/proton-antiproton/proton-ion/ion-ion reactions, a significant number of beam particles disintegrates in each bunch-to-bunch collision, creating an entangled picture of events in the particle detectors which cannot be temporally resolved:

$$N_{pileup} = \frac{L\sigma_{inel}}{f_0 N_B}, \tag{1.9}$$

where the inelastic events cross section is approximately σ_{inel} [mbarn] $\approx 70 + 20 \log$ (E[TeV]) [31]. To keep the number of events per bunch crossing sufficiently low under the condition of high average luminosity, one needs to increase the number of bunches—indeed, in the Tevatron the pile-up was kept under $N_{pileup} < 15$ with $N_B = 36$ bunches, while the pileup of up to 30 is expected in the LHC at the design luminosity with $N_B = 2808$ bunches.

Separation of two beams is needed to avoid multiple collisions points which would immediately lead to unacceptable total beam-beam tuneshift parameters ξ—see (1.8)—and leave only one or a few dedicated IPs with head-on collisions. Such separation can be implemented either by the use of high-voltage (HV) electrostatic separators in single-aperture proton-antiproton colliders, like Tevatron, or by an independent aperture for each beam, like in RHIC (two magnetic systems) or in the LHC ("two-in-one" magnet with two beam openings, each having same but opposite magnetic fields). But even in the latter case, there are, by necessity, regions near the main IPs where two beams have to join each other in

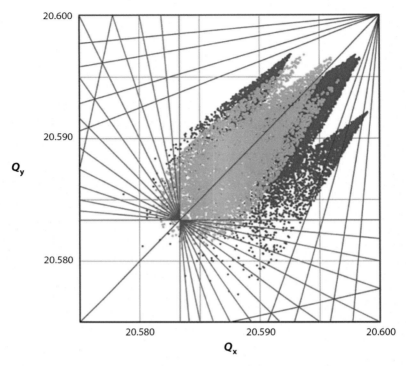

Fig. 1.9 Tevatron proton and antiproton tune distributions superimposed onto a resonance line plot. The *red* and *green lines* are various sum and difference tune resonances of up to the 12th order. The *blue dots* represent the calculated tune distributions for all 36 antiproton bunches; the *yellow* represent the protons. The tune spread for each bunch is calculated for particles up to 6σ amplitude taking into account the *measured* intensities and emittances (from [30])

the same vacuum chamber. Usually these regions are much longer than the bunch-to-bunch spacing (longitudinal distance between neighboring bunches) and a significant number of such parasitic *long-range beam-beam* interactions between separated bunches can take place. For example, there are 70 long-range encounters per revolution in the Tevatron and up to 120 in the LHC. Altogether they can produce significant, sometimes dominating effects on the beam dynamics. It is typical for hadron colliders that beams are brought to head-on collisions only at the very end of the preparatory part of the operational HEP store cycle. Therefore, the beam dynamics during preceding stages of injection, energy ramp, and the low-beta squeeze (focusing optics adjustment) is mostly driven by the long-range beam–beam effects. The parasitic interaction effects depend on the normalized radial separation S of the beams orbits [27, 32, 33], expressed in the units of the RMS betatron beam sizes $\sigma_{x,y\beta}$:

$$S = \sqrt{\left(\Delta x/\sigma_{x\beta}\right)^2 + \left(\Delta y/\sigma_{y\beta}\right)^2}. \tag{1.10}$$

Experience in the Tevatron and the LHC indicates that even a few long-range encounters with $S < 5$–6 could causes unsatisfactory losses. Hence, the supercollider designers typically aim at the orbit separations of $S \sim 9$ or so. The long-range interactions contribute a tune spread of about:

$$\Delta Q_{LR} \approx \sum_{parasitic\ encounters} \frac{2\xi_{IP}}{S^2}. \qquad (1.11)$$

The pattern of the long-range interactions and, consequently, their strength, differs for each bunch with especially large variations at the end of bunch trains. This makes all beam dynamics indicators dependent on the bunch position in the train of bunches. For example, in the Tevatron collider operating with three trains of 12 bunches in each beam, there were observed: significant bunch-to-bunch variations of the beam orbits of about 40 µm, of the betatron tunes—by as much $\Delta Q_{LR} \approx 0.005$, of the coupling and of the chromaticities $\Delta Q'_{LR} = \Delta(dQ/(dp/p))_{LR} \approx 6$ [21]. It is not surprising that with such significant differences in the tunes and chromaticities, the antiproton and proton bunch intensity lifetime and emittance growth rates varied substantially from bunch to bunch (we will discuss that in more detail discussion in the following sections).

Other complications of the beam-beam interactions can come from the facts that bunch dimensions at the IPs are not always the same for two opposing beams or in the vertical and horizontal planes, or when intensities of colliding beams are significantly mismatched. Many other factors—like the presence of either external noises or machine impedance, synchrotron radiation damping or other means of beam cooling, nonlinear magnetic focusing components in the machine or collisions assisted by crab-cavities or crab-waist elements—can play a significant role, too. All in all, the beam-beam effects remain one of the most critical challenges for hadron supercolliders.

Let us use the Tevatron collider as an illustration of the importance of the beam-beam interactions. During the Collider Run II (2001–2011), the beam losses during injection, ramp and low-beta squeeze stages were mostly caused by beam-beam effects. Figure 1.10 shows that early in the Run II, combined beam losses only in the Tevatron itself (the last accelerator out of total 7 in the accelerator chain) claimed significantly more than half of the luminosity. Due to various improvements, the losses have been reduced significantly down by some 20–30 % in 2008–2009, paving the road to a many-fold increase of the luminosity. In "proton-only" or "antiproton-only" stores, i.e., without any collisions, the losses did not exceed 2–3 % per species. So, the remaining 8–10 % proton loss and 2–3 % antiproton loss at the opertational stages preceding the start of the HEP store were due to the beam-beam effects, and they correspondingly reduced the initial luminosity L_0.

During the HEP stores, the Tevatron luminosity decay could be well approximated by a simple empirical fit with just two parameters—the initial luminosity L_0 and the initial luminosity lifetime τ_L [34]:

$$L(t) = \frac{L_0}{1 + t/\tau_L}, \qquad (1.12)$$

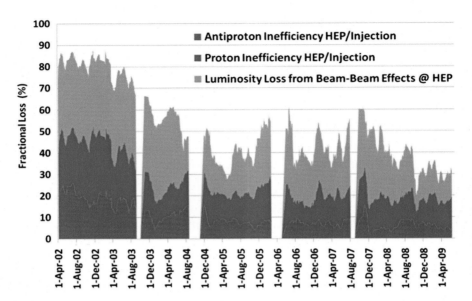

Fig. 1.10 Evolution of the Tevatron beam losses in 2002–2009. *Red* shows fractional loss of antiprotons between injection into the Tevatron and start of collisions, next (*blue*) shows loss of protons, *green*—fractional reduction of the luminosity integral caused by beam-beam effects in collisions [34]

Corresponding luminosity integral for the entire store duration T scales as the product of the peak luminosity and the luminosity lifetime $I \approx L_0 \tau_L \ln(1 + T/\tau_L)$. According to (1.6), the collider luminosity lifetime is depends on the speed of the emittance growth, proton and antiproton beam intensity loss rates, and bunch lengthening (the latter affects the geometric luminosity reduction factor F):

$$\tau_L^{-1} = \frac{dL(t)}{L(t)dt} = \left| \tau_\varepsilon^{-1} \right| + \tau_{Na}^{-1} + \tau_{Np}^{-1} + \tau_F^{-1}. \qquad (1.13)$$

At the end of Run II, the luminosity loss rates were in the range of 19–21 %/h at the beginning of stores, i.e., $\tau_L \approx 5$ h. For the 2010–2011 HEP stores in range of initial luminosities were between 3.0 and 4.3 10^{32} cm^{-2} s^{-1}, and the largest contribution to luminosity decay came from beam emittance growth with a typical time of $\tau_\varepsilon \sim 9$–11 h. The growth was dominated by intrabeam scattering (IBS) in the proton bunches, with small contributions from the IBS in antiprotons and external noises. Beam-beam effects usually manifested themselves in reduction of the beam emittances or their growth rates. The antiproton bunch intensity lifetime $\tau_a \sim 16$–18 h was dominated by the luminosity burn-up with a total cross section of about σ_{tot} [mbarn] ≈ 70 mbarn—see (1.5)—which accounted for 80–85 % of the lifetime, while the remaining 10–15 % came from parasitic beam-beam interactions with protons. Proton intensity loss varied in a wide range $\tau_p \sim 25$–45 h and was driven

mostly (~50 %) by the head-on beam-beam interactions with smaller size antiprotons at the main IPs. The proton lifetime caused by inelastic interactions with
antiprotons in collisions and with residual gas molecules did vary from 300 to
400 h. The hourglass factor F decayed with $\tau_F \sim 70\text{–}80$ h due to the IBS, again,
mostly in proton bunches. Beam-beam effects sometimes lead to reduction of the
proton bunch length growth (longitudinal "shaving") in a poorly tuned machine.
Combining all of these loss rates together, one can estimate the hit on the luminosity
lifetime τ_L due to the beam-beam effects can be as much as 12–17 % (that is equal to
2.5–3.5 %/h out of total 19–21 %/h). Therefore, the full impact of the beam-beam
effects on the luminosity integral should include beam-beam driven proton and
antiproton losses at the injection energy (about 5 and 1 %, correspondingly), on
the energy ramp (2 and 3 %), and in the low-beta squeeze (1–2 % and 0.5 %) which
proportionally reduce the initial luminosity L_0. So, altogether, at the end of the
Tevatron collider Run II operation, after about a decade of studies, improvements
and optimizations, the beam-beam effects resulted in overall reduction of the luminosity integral by 23–33 % [21].

   Operational implications of the beam-beam interactions in RHIC and LHC are
similarly serious [35] and will be presented and considered in more detail in the
following chapters.

1.2.2   Single Beam Issues in Hadron Colliders

Besides the issues related to beam-beam interactions, many challenges arise from
one-beam phenomena specific for hadron supercolliders and driven by the quest for
higher luminosity. Comprehensive analysis of them is beyond the scope of this
book so we only briefly overview them. An interested reader can be referred to
detailed reviews in, e.g., [13, 21, 36, 37]. Taking into account the beam-beam
tuneshift limit (1.8), the luminosity equation (1.6) can be re-written as:

$$L = \xi \cdot \frac{I}{e} \cdot \frac{\gamma}{\beta^*} \cdot \frac{F}{r_p}, \qquad (1.14)$$

where $I = efN_BN_p$ is the single beam total current, $\gamma = E/mc^2$ is relativistic factor
and e is the proton charge. The form factor F is always tried to be set as close to 1 as
possible, so the focus of the luminosity maximization efforts – under the condition
of certain beam-beam limit on parameter ξ – usually goes towards an increase of the
total beam current I and reduction of the beta-function β^* at the IPs.

   The issues associated with high-current operation are numerous and start with
the beam production. One of the most severe limitations is set by space-charge
forces in the low-energy accelerators in the collider injection chain which lead to
unacceptable emittance growth and particle losses if the space-charge tuneshift
parameter:

$$-\Delta Q_{SC} = \frac{N_p r_p B_f}{4\pi\varepsilon_n \beta_p \gamma^2}, \qquad (1.15)$$

exceeds a certain value, typically about ~0.3 (here $\beta_p = v_p/c$ and γ are relativistic parameters, B_f is the bunching factor—ratio of the peak current to average current) [38, 39]. Production and accumulation of secondary particles, such as antiprotons has its own set of challenges associated with the need of the beam cooling [21, 40]. High current proton beams are subject to various collective instabilities both in the injectors and collider rings caused by the machine impedance [41] and electron-cloud initiated by either particle losses or by synchrotron radiation [42]. Various cures are being implemented to keep such high intensity beams stable, including minimization of the impedance, feedback systems to damp beam oscillations, introduction of non-linear focusing elements to induce stabilizing tune spread in the beams, operation at higher chromaticities $Q' = dQ/(dp/p)$, special coating or beam scrubbing of the beam vacuum chambers by enhanced proton losses to suppress formation of the electron cloud, etc.

One of the biggest challenges for energy frontier superconducting colliders is control and safe absorption of ever growing energy $W = IE/f_0$ carried by the beams—see Table 1.2—and beam backgrounds. For example, in the LHC, with $I = 0.5$ A of circulating current and at $E = 7$ TeV beam energy, the stored energy is $W = 360$ MJ per beam in the nominal conditions—(see Fig. 1.11a). Even modest 1 % loss of the stored beam intensity over a period of about 10 s would produce a peak power load of 500 kW. That surely would pose a high risk as the quench limit in the main LHC magnet is about 8 W/m. Sophisticated collimation systems are required to protect the machine and the HEP detectors from beam losses and particle halo backgrounds. A two-stage collimation system of the Tevatron collider comprised of 14 collimators and successfully handled some 2 MJ of beam energy at 1 TeV [44], while 108 collimators were installed in the LHC three-stage collimation system to guarantee inefficiencies of less or about 2×10^{-5} per m of accelerator for a circulating current of 3×10^{14} protons per beam [45]—see Fig. 1.12. To assure

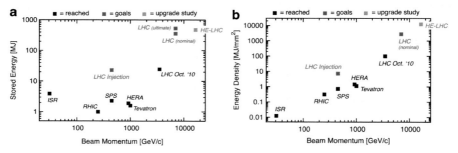

Fig. 1.11 (**a**—*left*) Stored energy per beam vs. beam momentum for record high energy, high intensity accelerators. The *black squares* indicate achieved values, *red squares* show design values and the *blue square* represents the HE-LHC design [37]. (**b**—*right*) Energy density vs. beam momentum (from [43])

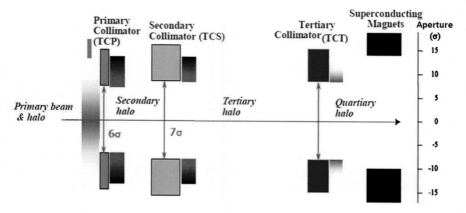

Fig. 1.12 Schematic diagram of the LHC three-stage collimation system: the primary collimator jaws are positioned closest to the beam while the jaws of the secondary and tertiary collimators are retracted further away from the beam (from [45])

minimal inefficiency—defined as unwanted fractional leakage of high energy particles from the collimation system onto critical machine elements, such as SC magnets—the collimators need to be placed within fractions of a mm to the beam and become very troublesome dominating sources of impedance which in turn can limit the maximum machine current [46]. Due to smaller beam size and higher beam energy, record high energy densities in the LHC and future hadron supercolliders—presented in Fig. 1.11b—pose serious concerns about the collimator material robustness against regular beam losses and accident scenarios [45].

Stringent requirements on the beam loss in supercolliders put additional emphasis on the beam orbit control and ground motion drifts [47], intrabeam and gas scattering, external noises [21], single particle diffusion due to nonlinear resonances, etc. Careful polarization control in RHIC [48] and additional challenges of ion-ion or proton-ion collisions in RHIC and LHC [49] also add to the long list of issues which superconducting supercolliders face.

1.3 Overview of the Electron Lens Technique and Its Applications in Supercolliders

Electron lenses are low-energy, magnetically confined electron beams whose electromagnetic fields are used for active manipulation of the circulating beams in high-energy accelerators. Assuming, for simplicity, axially symmetric electron current-density distribution $j_e(r)$ one can calculate the force exerted by electron lens on the relativistic proton going through it parallel to its axis:

$$F_r(r,t) = e\left(E_r + \beta_p B_\theta\right) = \left(1 \mp \beta_p \beta_e\right) \frac{4\pi e}{c\beta_e} \left(\frac{1}{r} \int\limits_0^r j_e(r,t)r\,dr\right). \qquad (1.16)$$

There "$-$" sign in the first factor corresponds to the case when proton and electron velocities $v_{e,p} = c\beta_{e,p}$ being parallel, and "$+$" to antiparallel case. For example, a 3 A 1 mm diameter beam of 10 kV electrons generates ~1 MV/m fields. Electron space charge forces are linear at distances smaller than the characteristic beam radius $r < a_e$ but scale as $1/r$ for $r > a_e$. Immersion of the electron beam in strong longitudinal magnetic field allows generation of transversely very stable, high current, narrow beams of a variety of desired density profiles $j_e(r)$ from external electron sources. There are a number of advantages which promote application of electron lenses in high energy hadron colliders:

(a) The electron beam acts on high-energy beams only through electro-mag-
      netic (EM) forces, with no nuclear interactions: $e-p$ cross sections even for
      multi-TeV protons colliding with non-relativistic electrons are thousands of
      times smaller than p-p cross sections of interactions with residual gas [31, 50,
      51], so, the typical electron densities in e-lenses of about $n_e \sim 10^{11}$ cm^{-3} are
      equivalent to miniscule additional gas pressure of less than 10^{-14} Torr over
      just few meters of collider circumference; therefore, electron lenses, while
      acting on high-energy protons, do not present any material close to the beam,
      thus, avoiding material damage and impedance increase;
(b) In a typical e-lens configuration, electrons are produced at the cathode of an
      electron gun and right after interaction with (anti) protons they are damped in a
      collector; thus, fresh electrons interact with the high-energy particles on each
      turn, leaving no possibility for coherent instabilities, similar to those that are
      driven by the electron-cloud or those which prevented a four-beam beam-
      beam compensation in the DCI experiment in 1970s [52];
(c) The electron current profile $j_e(r)$ and, thus, the EM field profiles can be easily
      changed for different applications—flat, Gaussian, hollow, etc.—by shaping
      the cathode and the extraction electrodes in the electron gun and by adjusting
      their relative electric potentials;
(d) The electron beam current can be varied quickly on very fast time scales O
      (10 ns) and e-lenses in pulsed operation, enabled by the availability of high
      voltage modulators with fast rise times, can therefore be synchronized with
      individual proton bunches or subsets of bunches with different intensities for
      each subset.

    One has to note, that the use of the space-charge forces of accumulated electrons is not particularly a new idea, and practical systems, called "Gabor lenses" or "plasma lenses" were developed long ago and used for focusing of proton and ion beams—see, e.g., [53–55]. They employ the charge and current screening effects in pre-arranged sections of dense plasma. But exactly due to the above listed reasons, these systems are exclusively used only in single-pass beam accelerators—namely, they cannot be used in the multi-turn machines like colliders because the presence of the plasma ions leads to fast loss of circulating proton/ion beams (with associated concerns for the particle detector backgrounds and radiation damage of the accelerator components), all kinds of plasma instabilities pose serious concerns for beams which are supposed to circulate for millions or even billions of turns, careful control of the plasma charge profiles is very challenging and plasma density modulation at the required time scales is problematic. Comprehensive theoretical analysis of the possible employment of the plasma screening in various lepton and hadron colliders concludes that such systems are feasible only for the future high energy muon colliders [56].

The electron lenses for supercolliders were originally considered in 1993 by Tsyganov et al. [57] to reduce the tune spread due to beam-beam interactions in the SSC and, independently, in 1997 Shiltsev et al. proposed them for compensation of the long-range and head-on beam-beam effects in the Tevatron proton-antiproton collider [58, 59]—see Figs. 1.13 and 1.14.

Comprehensive theory of the electron lens beam-beam compensation and detail design considerations were accomplished by 1999 [60]. The first two electron lenses for beam-beam compensation were developed, built and installed in the Tevatron in 2001 and 2004 [61]—see Fig. 1.15. They enabled the first observation of long-range beam-beam compensation effects by tune shifting individual bunches [62] and experiments on head-on beam-beam compensation in 2003 and 2009 [30, 63]. Since 2001 until the end of the Tevatron Collider Run II in 2011, the TELs were used during regular Tevatron collider operations for longitudinal beam collimation—removal of uncaptured protons and antiprotons from the abort gaps [64]. Transverse halo collimation by electron lenses was proposed in 2006 [65] and

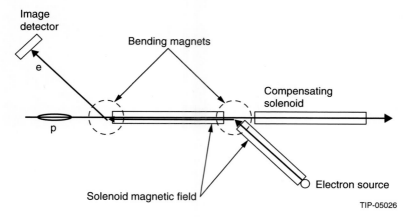

Fig. 1.13 Schematics of the proposed beam-beam compensating device for the SSC. It was motivated by the need to increase the transverse decoherence time be reduction of the betatron tune spread and, thus, to ease the requirements on the beam feedback system for emittance preservation. Low-energy electrons beam collide with a bunch of protons but are kept stable in space by a solenoidal magnetic field. After collision with the proton bunch, electron beam is deflected to the image detector, which is used for steering onto the proton bunch (from [57])

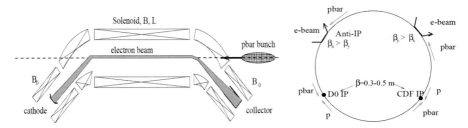

Fig. 1.14 (**a**—*left*) Schematic layout of the Tevatron electron lens for head-on beam-beam compensation (from [58]); (**b**—*right*) two electron lenses with specially preset pulsed current waveforms, installed in the locations with un-equal vertical and horizontal beta-functions, can compensate bunch-by-bunch tune spread caused by long-range beam-beam interactions in the Tevatron. The Tevatron e-lenses (TELs) were expected to increase collider luminosity by allowing higher proton bunch intensities and more bunches (from [59])

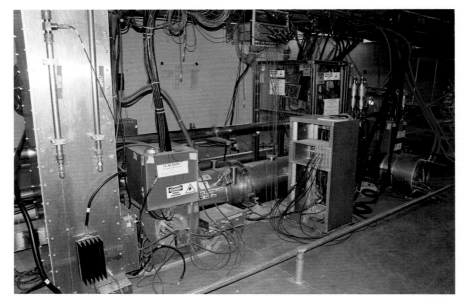

Fig. 1.15 The first electron lens system, installed in the Tevatron proton-antiproton collider and operational since 2001. Subsystems (from *left* to *right*): anode high voltage modulator; electron gun inside the gun solenoid, main superconducting solenoid, rack of the TEL BPM electronics, the collector solenoid and the electron collector

one of the two TELs was used for successful demonstration of the hollow electron beam collimation in 2010–2011 [66]. Electron lenses for head-on beam-beam compensation in RHIC at BNL were proposed in 2007 [67], two of them have been built, installed and commissioned [68] and allowed doubling the collider luminosity in 2015 [69]. Electron-lens compensation of space-charge effects in high-intensity proton accelerators, including super-collider's injectors, was proposed in 2000 [70] and is now subject of the experimental accelerator research program at the IOTA test ring at Fermilab [71].

Unique advantages of the electron lens technique have triggered a number of proposals for many other applications, such as, e.g., arrangements towards nonlinear integrable lattices [72, 73] (1997–1998), selective slow extraction system [74] (2001), beam-beam compensation in $e + e-$ colliders [75] (2001) and electron-ion colliders [76] (2009), tune-spread generators for Landau damping of instabilities before collisions [65] (2006); and have instigated design considerations for the LHC upgrades such as halo scrapers [77, 78] (2007), as current-carrying 'beam-wires' for long-range beam-beam compensation [79, 80] (2007), and as "beam-beam" kickers [81, 82] (1996).

Similar to the figures of merit for beam-beam interactions (1.8) and for the space-charge effects (1.15), the shift of the particle's (e.g., proton's) betatron frequencies by electron beam $dQ_{x,y}$ can be used as the figure of merit for the electron lenses in supercolliders:

$$dQ_{x,y} = \frac{\beta_{x,y} L_e r_p}{2\gamma ec} \cdot j_e \cdot \left(\frac{1 \mp \beta_e}{\beta_e} \right), \tag{1.17}$$

where $\beta_{x,y}$ are the Courant-Snyder beta-functions (horizontal, vertical) at the locations of the lenses, L_e and j_e are the length and current density of the electron beam—see (1.16). For many applications of the electron lenses in the hadron supercolliders, the required tune shift $dQ_{x,y}$ is of the order of 0.01.

To conclude this chapter—the electron lenses present a novel instrument for high energy particle accelerators, their flexibility allows many applications focused on the luminosity improvements in the energy-frontier supercolliders and other high-intensity proton accelerators.

References

1. E. Haussecker, A. Chao, Phys. Perspective **13**, 146 (2011)
2. M. Tigner, A. Chao (eds.), *Handbook of Accelerator Physics and Engineering* (World Scientific, Singapore, 1999)
3. V. Shiltsev, Phys.-Uspekhi **55**(10), 965 (2012)
4. M. Tigner, Nuovo Cimento **37**, 1228 (1965)
5. S. Holmes, V. Shiltsev, Annu. Rev. Nucl. Part. Sci. **63**, 435–465 (2013)
6. L. Evans, Annu. Rev. Nucl. Part. Sci. **61**, 435–466 (2011)
7. A. Tollestrup, E. Todesco, Rev. Accel. Sci. Tech. **1**, 185–210 (2008)
8. M. Harrison, S. Peggs, T. Roser, Annu. Rev. Nucl. Part. Sci. **52**(1), 425–469 (2002)
9. *SSC Conceptual Design,* SSC-SR-2020, SSC Laboratory (1986)
10. O. Bruning, P. Collier, Nature **448**, 285–289 (2007)
11. V. Yarba, in *Proceedings of IEEE PAC '91* (San Francisco, CA, USA, 1991), p. 2013

12. S. Wojcicki, Rev. Accel. Sci. Tech. **2**, 265 (2009)
13. VLHC Design Study Group, *Design study for a staged very large hadron collider*, FERMILAB-TM-2149 (2001)
14. F. Zimmermann et al., in *Proceedings of IPAC'2014* (Dresden, Germany, 2014), p. 1
15. V. Shiltsev, JINST **9**, T07002 (2014)
16. L. Teng, FERMILAB-FN-461 (1987)
17. V. Yarba, private communication (2014)
18. L. Rossi, L. Bottura, Rev. Accel. Sci. Tech. **5**, 51–89 (2012)
19. Y. Yan, J. Naples, M. Syphers (eds.), *Accelerator Physics at the SuperConducting Super Collider* (AIP Conference Proceeding 326, AIP, New York, 1995)
20. S. Myers, H. Schopper (eds.), *Elementary Particles-Accelerators and Colliders,* Landolt-Bornstein: Numerical Data and Functional Relationships in Science and Technology-New Series/Elementary Particles, Nuclei and Atoms, **21** (Springer, Berlin, 2013)
21. V. Lebedev, V. Shiltsev (eds.), *Accelerator Physics at the Tevatron Collider* (Springer, New York, 2014)
22. W. Scandale, Rev. Accel. Sci. Tech. **7**, 9–33 (2014)
23. I.A. Koop, G.M. Tumaikin (Eds.), in *Proceedings of 3rd Advanced ICFA Beam Dynamics Workshop on Beam-Beam Effcets in Circular Colliders* (Novosibirsk, USSR, May 29–June 3, 1989)
24. J. Poole, F. Zimmermann (eds.), in *Proceedings of Workshop on Beam-Beam Effects in Large Hadron Colliders* (Geneva, April 12–17, 1999), CERN-SL-99-039 AP (1999)
25. J. Wei, W. Fischer, P. Manning (eds.), in *Proceedings of 29th ICFA Beam Dynamics Workshop HALO'03* and *Workshop on Beam-Beam Interactions BEAM-BEAM'03* (AIP Conference Proceeding 693, AIP, New York, 2003)
26. W. Herr, G. Papotti (eds.), in *Proceedings of ICFA Mini-Workshop on Beam-Beam Effects in Hadron Colliders* (18–22 March, 2013, CERN, Geneva), CERN-2014-004 (2014)
27. T. Sen et al., Phys. Rev. ST Accel. Beams **7**, 041001 (2004)
28. V. Shiltsev et al., Phys. Rev. ST Accel. Beams **8**, 101001 (2005)
29. W. Fischer (Ed.), ICFA Beam Dynamics Newsletter **52** (2010)
30. V. Shitsev et al., New J. Phys. **10**, 043042 (2008)
31. K. Olive et al., (Particle Data Group), Chin. Phys. C **38**, 090001 (2014)
32. D. Siergiej, D. Finley, W. Herr, Phys. Rev. E **55**, 3521 (1997)
33. Y. Papaphilippou, F. Zimmermann, Phys. Rev. ST Accel. Beams **5**, 074001 (2002)
34. R. Moore, A. Jansson, V. Shiltsev, JINST **4**, P12018 (2009)
35. see, e.g., Y. Luo, W. Fischer, in *Proceedings of ICFA Mini-Workshop on Beam-Beam Effects in Hadron Colliders*, ed. by W. Herr, G. Papotti (18–22 March, 2013, CERN, Geneva), CERN-2014-004 (2014), pp. 19–25; and G.Papotti et al, *ibid.*, pp.1–5
36. O. Brüning et al., *The LHC Design Report, CERN Technical Report* (CERN, Geneva, 2004) http://ab-div.web.cern.ch/ab-div/Publications/LHC-designreport.html
37. E. Todesco, F. Zimmermann (Eds.), in *Proceedings of EuCARD-AccNet-EuroLumi Workshop "The High Energy Large Hadron Collider"* (14–16 October, 2010, Malta), CERN-2011-003 (2011)
38. A. Luccio, W. Weng (Eds.), in *Proceedings of Workshop on Space Charge Physics in High Intensity Hadron Rings* (May 4–7, 1998, Shelter Island, NY), AIP Conference Proceeding 448 (AIP, New York, 1998)
39. See, e.g., presentations at the *"SPACE CHARGE 2013" Workshop* (16–19 April 2013, CERN) https://indico.cern.ch/event/221441/
40. M. Church, J. Marriner, Annu. Rev. Nucl. Part. Sci. **43**, 253–295 (1993)
41. K.Y. Ng, *Physics of Intensity Dependent Beam Instabilities* (World Scientific, Singapore, 2006)
42. R. Cimino, G. Rumolo, F. Zimmermann (eds.), in *Proceedings of ECLOUD'12: Joint INFN-CERN-EuCARD-AccNet Workshop on Electron-Cloud Effects* (5–9 Jun 2012, La Biodola, Italy), Preprint CERN-2013-002 (2013)

43. see R. Assmann, in *Proceedings of EuCARD-AccNet-EuroLumi Workshop "The High Energy Large Hadron Collider"*, ed. by E. Todesco, F. Zimmermann (14–16 October, 2010, Malta), CERN-2011-003 (2011), pp. 124–127
44. N. Mokhov et al., JINST **6**, T08005 (2011)
45. M. Cauchi et al., Phys. Rev. ST Accel. Beams **17**, 021004 (2014)
46. H. Burkhardt et al., in *Proceedings of IEEE PAC'05* (Knoxville, TN, USA, 2005), p. 1132
47. V. Shiltsev, Phys. Rev. ST Accel. Beams **13**, 094801 (2010)
48. M. Bai et al., in *Proceedings of IPAC'2013* (Shanghai, China, 2013), p. 1106
49. J. Jowett et al., in *Proceedings of IPAC'2013* (Shanghai, China, 2013), p. 49
50. M. Christy, P. Bosted, arxiv:0712.3734 (2013)
51. S. Striganov, private communication (2015)
52. Orsay Storage Ring Group, in *Proceedings of IEEE PAC'1979* (San Francisco, CA, USA, 1979), p. 3559
53. D. Gabor, Nature **160**, 89–90 (1947)
54. J. Palkovic, *Gabor lens focusing and emittance growth in a low-energy proton beam*, PhD Thesis, University of Wisconsin (1991). FERMILAB-THESIS-1993-64 (1993)
55. B.G. Logan et al., Nucl. Fusion **45**, 131 (2005)
56. G. Stupakov, P. Chen, Phys. Rev. Lett. **76**(20), 3715 (1996)
57. E. Tsyganov, A. Taratin, A. Zinchenko, Preprint SSCL-Report No. 519 (1993)
58. V. Shiltsev, D. Finley, Preprint FERMILAB-TM-2008 (1997)
59. V. Shiltsev, Preprint FERMILAB-TM-2031 (1997)
60. V. Shiltsev, V. Danilov, D. Finley, A. Sery, Phys. Rev. ST Accel. Beams **2**(7), 071001 (1999)
61. V. Shiltsev et al., Phys. Rev. ST Accel. Beams **11**, 103501 (2008)
62. V. Shiltsev et al., Phys. Rev. Lett. **99**, 244801 (2007)
63. G. Stancari, A. Valishev, in *Proceedings of ICFA Mini-Workshop on Beam-Beam Effects in Hadron Colliders*, ed. by W. Herr, G. Papotti (18–22 March, 2013, CERN, Geneva), CERN-2014-004 (2014), p. 121
64. X.-L. Zhang et al., Phys. Rev. ST Accel. Beams **11**, 051002 (2008)
65. V. Shiltsev, in *Proceedings of CARE-HHH-APD LHC-LUMI-06 Workshop "Towards a Roadmap for the Upgrade of the CERN & GSI Accelerator Complex"* (16–20 October 2006, Valencia), Yellow Report CERN-2007-002 (2007)
66. G. Stancari et al., Phys. Rev. Lett. **107**, 084802 (2011)
67. Y. Luo, W. Fischer, Report BNL C-A/AP/286 (2007)
68. W. Fischer et al., in *Proceedings of IPAC'2014* (Dresden, Germany, 2014), p. 913
69. V. Schoefer et al., in *Proceedings of IPAC'2015* (Richmond, VA, USA, 2015), p. 2384
70. A. Burov, G. Foster, V. Shiltsev, Fermilab Preprint FNAL-TM-2125 (2000)
71. V. Shiltsev, M. Chung, Fermilab Preprint CONF-14-324-APC (2014)
72. V. Danilov, E. Pervedentsev, in *Proceedings of IEEE PAC'1997* (Vancouver, Canada, 1997), p. 1759
73. V. Danilov, V. Shiltsev, Preprint FERMILAB-FN-0671 (1998)
74. V. Shiltsev, J. Marriner, in *Proceedings of IEEE PAC'2001* (Chicago, IL, USA, 2001), p. 1468
75. V. Shiltsev, in *Proceedings of 23rd Advanced ICFA Beam Dynamics Workshop on High Luminosity e + e−Colliders "FACTORIES-2001"* (October 15–19, 2001, Ithaca NY, USA). http://www.lns.cornell.edu/public/icfa/proceedings/index.html
76. C. Montag, W. Fischer, Phys. Rev. ST Accel. Beams **12**, 084001 (2009)
77. V. Shiltsev, in *Proceedings of the CARE-HHH-APD Workshop on Finalizing the Roadmap for the Upgrade of the CERN and GSI Accelerator Complex BEAM07 (CERN, Geneva, Switzerland, October 2007)*, CERN-2008-005 (2008), p. 46
78. G. Stancari et al., FERMILAB-TM-2572-APC (2014)
79. V. Shiltsev, in *Proceedings of the CARE-HHH-APD Workshop on Finalizing the Roadmap for the Upgrade of the CERN and GSI Accelerator Complex BEAM07 (CERN, Geneva, October 2007)*, CERN-2008-005 (2008), p. 16
80. A. Valishev, G. Stancari, FERMILAB-TM-2571-APC (2013)

81. V. Shiltsev, Nucl. Instrum. Meth. A **374**(2), 137–143 (1996)
82. Y. Zhang, J. Bisognano (eds.), *Science Requirements and Conceptual Design for a Polarized Medium Energy Electron-Ion Collider at Jefferson Lab,* arxiv:1209.0757 (2012)
83. see, e.g., G. Papotti et al., in *Proceedings of ICFA Mini-Workshop on Beam-Beam Effects in Hadron Colliders*, ed. by W. Herr, G. Papotti (18–22 March, 2013, CERN, Geneva), CERN-2014-004 (2014), pp. 1–5

Chapter 2
Technology of Electron Lenses

2.1 Major Requirements

Table 2.1 below presents major requirements of the electron lenses for various applications head-on beam-beam compensation [1–10], long-range beam-beam compensation [11, 12] hollow electron beam collimation [13, 14] and space-charge compensation [15]. The needed electron beam currents vary from 0.3 to 20 A, electron beam energies are in the range of 5–80 kV and beam sizes in the range from 100 μm to 3.5 mm with typical length of the lenses being about 2–4 m. Current modulation requirements vary from 100 ns to DC operation.

These requirements are quite challenging to meet because of the strong space change forces of intense electron beams. Indeed, in a free space, the electron beam will expand and lose its shape and density over the distance of about:

$$z_L \approx \frac{a_e}{\sqrt{K}}.$$

(2.1)

where a_e is beam radius and the generalized beam perveance K scales with beam current J_e as [11]:

$$K \approx \frac{J_e}{I_0} \cdot \frac{2}{\beta_e^3 \gamma_e^3},$$

(2.2)

$I_0 = 17$ kA is the characteristic current for electrons. For a 3 A 10 kV electrons, $K \approx 0.044$ and a 2 mm diameter beam will expand just over 1 cm. The preferred solution for this problem in the electron lenses is to immerse the electron beam in a strong longitudinal magnetic field. Having a strongly magnetized electron beam (a) helps to keep the beam stable against its own space charge forces, (b) suppresses the distortions due to electromagnetic forces of high energy (anti)proton bunches passing through the electron beam, and (c) assures sufficient transverse rigidity of

© Springer Science+Business Media New York 2016
V.D. Shiltsev, *Electron Lenses for Super-Colliders*, Particle Acceleration and Detection, DOI 10.1007/978-1-4939-3317-4_2

Table 2.1 Parameters of the electron lenses for head-on and long-range beam-beam compensation (HO-BBC, LR-BBC), hollow electron beam collimation (HEBC), space-charge compensation (SCC): shape of the electron current density profile—Gaussian, hollow or "smooth-edges-flat-top" (SEFT), electron beam current J_e, energy U_e, size a_e, length L_e and characteristic modulation time τ_e (bunch spacing, revolution period, etc.)

		Profile	J_e (A)	U_e (kV)	a_e (mm)	L_e (m)	τ_e (μs)
HO-BBC	SSC	Gaussian	0.03	10	0.1	2.0	DC
	TEV		1	10	0.7	2.0	DC/0.4
	RHIC		1.1	10	0.3	2.1	DC/0.1
	LHC		1.2	10	0.3	3.0	DC
LR-BBC	TEV	SEFT	0.6–3	10	2.3	2.0	0.4/0.13
	LHC		20	10–20	2	3.0	0.4
HEBC	TEV	Hollow	1	5	3	2.0	DC
	LHC		2–10	10–20	2.5	3.0	DC/0.2
SCC	Booster	Gaussian	12	80	3.5	4.0	0.02/AC

the electrons to avoid any coherent instabilities in the high-energy beams. Besides being used for the transport of electrons from the source (a cathode) to the interaction region and then to the absorber (collector), the magnetic system of the electron lens allows precise positioning of the electron beam on the high-energy beam of choice and to control the electron beam size in the interaction region by changing adiabatic magnetic compression factor. Indeed, due to the conservation of the adiabatic invariant $a^2_e B = const$, the cross-section area of a strongly magnetized electron beam following the magnetic field lines scales with the magnetic field B as $1/B$, therefore, changing the ratio of the fields in the interaction region and at the electron gun cathode B_{main}/B_{gun} allows compression of the beam size and the current density. The compression factor can be as big as 10 or so in a typical system with $B_{gun} = 0.2$–0.4 T and $B_{main} = 2$–6 T.

2.2   General Considerations and Specifications on the Electron Lens Subsystems

As mentioned above, a typical electron lens employs a low energy (dozen kV), high current (few A) electron beam interacting with protons or antiprotons in a strong magnetic field of several Tesla. The electron source is usually a thermionic oxide cathode of an electron gun immersed in somewhat weaker magnetic field. Electrons undergo adiabatic beam size compression which allows for an increase in the current density when they enter a few meters long straight interaction region with a stronger solenoid magnetic field. At the setup exit, the beam follows the magnetic field lines directed to the collector. In general, such a configuration is similar to low energy "electron cooling" devices [17, 18]. Our consideration below largely follows [3].

2.2.1   Electron Beam Considerations

In this section we consider the electron beam itself, and start with the choice of electron energy. The electron energy $U_e = m\beta_e^2 c^2/2$, or equivalently, the electron velocity β_e, is a trade-off between several factors. Some of them, such as space-charge potential, drift ion instability, current modulation time, etc., prefer higher U_e and β_e. Others, e.g., power of the electron gun and the modulator power supply, adiabatic compression, and total beam current, prefer smaller beam energy. Let us start with the general requirements on the electron beam for the beam-beam compensation. First, its size a_e must be either about the same as or just a few times the rms proton beam size and is typically of the order of 1 mm—see Table 2.1. The desired transverse current distribution depends on the electron lens application. In several numerical examples below we will, for simplicity, consider an electron beam with a radius of $a_e = 1$ mm and constant transverse distribution. If we aim at the certain figure of merit of the electron lens—the induced proton tune shift dQ (1.17)—then the electron beam current J_e scales with electron velocity as:

$$J_e = J_0 \frac{\beta_e}{1 + \beta_e}. \tag{2.3}$$

where the constant J_0 is about 10 A for most of the electron lens configurations listed in Table 2.1. The maximum current of a space charge limited diode electron gun is given by the Child-Langmuir law (see, e.g., [19])

$$J_e = PU_a^{3/2}. \tag{2.4}$$

where the perveance P is gun geometry dependent parameter, and U_a is the voltage difference between the cathode and the anode electrodes of the gun. Usually, perveance is presented in units of microperveance $P = \mu P \times 10^6$ A/V$^{3/2}$. In the case $U_e = U_a$ and $\beta_e << 1$, combining (2.3) and (2.4) one gets a minimum electron energy of

$$U_e \approx \frac{1.2 J_0}{P\sqrt{mc^2}} = \frac{16.5[kV]}{\mu P}. \tag{2.5}$$

The corresponding beam current and power are (for the same $J_0 = 10$ A):

$$J_e \approx \frac{2.1\ [A]}{\sqrt{\mu P}}, \quad W = J_e U_e = \frac{34\ [kW]}{\mu P^{3/2}}. \tag{2.6}$$

The energy recirculation technique widely used in electron cooling devices [18] allows for a reduction in the power dissipated on the electron collector. However, the high voltage power supply still has to provide power proportional to the total current $W_c = J_e U_c$, where $U_c = 1 \ldots 2$ kV is the cathode-collector potential difference (see also the next section). Therefore, since it is beneficial to reduce the beam current and power, a high gun perveance is needed. Higher perveance also helps to reduce the high voltage anode-cathode modulator power needed for the pulsed electron lens applications. For a diode gun with a flat cathode and a Pierce electrode, the microperveance is equal to [19] $\mu P = 7.3 \; (a/d)^2$, where a is the beam (cathode) radius and d is the cathode-anode distance. A rule of thumb is that good current density uniformity can be reached if the ratio of (a/d) is less than $1/2$, i.e., μP is less than 2. Several times higher perveance (up to 10) with good beam quality can be achieved with the use of a convex cathode immersed in a magnetic field of the order of 1 kG or more [20]. Such arrangement was found quite appropriate for the Tevatron and RHIC electron lenses which employ convex cathode electron guns with μP in the range of 1–6 (see below in Sect. 2.3.2).

2.2.1.1   Lower Limit on the Electron Beam Energy

The lower limit on the electron energy is set by two effects. First, the electrons must be fast enough to provide the necessary current modulation required for, e.g., long-range beam-beam compensation, and second, the electron kinetic energy must be sufficient to overcome the electron space-charge potential in a round vacuum chamber.

Let us consider the time structure of the electromagnetic kick (or the tune shift) produced by the electron lens. Figure 2.1 illustrates the effect of a simplified step-like current modulation with pulse duration of t_p (presented in the upper plot) on the antiproton bunches in the Tevatron. Let us denote t_0 the moment when the front of the electron pulse enters the interaction section. As the antiproton beam passes through the oncoming electron current pulse, the maximum deflection will be seen by test particles (antiprotons) which at t_0 are distanced by $(1 + \beta_e)L/\beta_e$ from the entrance of the device. We define the corresponding time $\tau_g = (1 + \beta_e)L/c\beta_e$ as the "kick growth time." The maximum kick lasts over a time interval of $t_f = t_p - \tau_g$ which is synchronized with the bunch arrival (see lower diagram in Fig. 2.1). Behind that bunch, the kick amplitude vanishes over the same "growth time". Let the required flattop of the kick be t_f, and the required "no-impact time" to be t_n, then, summarizing all time intervals in Fig. 2.1, for (anti)proton bunch spacing T_B one gets $t_f + 2(1 + \beta_e)L/c\beta_e + t_n < 2T_B$. The shorter the bunch spacing, the higher electron velocity is needed. That poses a stringent requirement for short bunch spacing applications, e.g., in the LHC with $T_B = 25$ ns the bunch-by-bunch tune control by a 2 m long electron lens would call for $\beta_e > 0.4$, i.e., $U_e > 40$ kV.

The other limit on the minimum voltage (kinetic energy) is set by the electron beam space-charge potential U_{sc} with respect to the grounded vacuum chamber walls

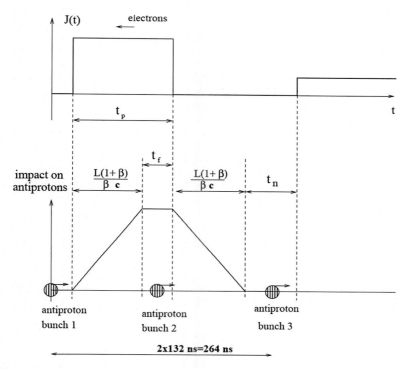

Fig. 2.1 Electron current pulse and its impact on high energy beam [3]

$$U_{sc} = \frac{2J_e}{\beta_e c} \left[\ln\left(\frac{b}{a_e}\right) + \frac{1}{2} \right], \tag{2.7}$$

where b is the beam pipe radius. In a typical case of $a_e \ll b$, the electron beam stability requires $U_e > 3/2 U_{sc}$ [21], e.g., higher than 3 kV for a 1 mm radius 2 A beam in 80 mm diameter chamber.

2.2.1.2 Electron Beam in Magnetic Field

The electron lens setup needs to have a longitudinal magnetic field in order to keep the electron beam envelope stable, make the beam more rigid, transport it from the cathode through the interaction region to the collector, and obtain a smaller beam size. The equation for paraxial electron oscillations under the impact of a solenoidal field B, the space-charge force due to the electron beam, and the force due to incoming (anti)protons is

$$\frac{d^2r}{ds^2} + r\left(\frac{1}{F_B^2} - \frac{1}{F_e^2} \pm \frac{1}{F_{p,a}^2}\right) = 0, \tag{2.8}$$

where s is the longitudinal coordinate along the electron orbit, and $r = |x + iy|$ is the oscillation amplitude (see [22, 23] for a detailed analysis of electron flows). The effective focal length due to the magnetic field B is

$$F_B = \frac{2\gamma_e\beta_e m_e c^2}{eB} = 0.33\ [\text{cm}]\ \frac{\gamma_e\beta_e}{B[T]}, \tag{2.9}$$

which for 10 kV electrons yields $F_B = 0.066$ [cm]/B[T]. It has to be compared with the defocusing length due to electron space charge which is given by

$$F_e = \sqrt{\frac{\gamma_e^3\beta_e^3 m_e c^3 a_e^2}{2eJ_e}} \approx 2.9[\text{cm}]\ \gamma_e^{3/2}\beta_e\sqrt{1+\beta_e}, \tag{2.10}$$

where we use (2.3) for the required electron current with $J_0 = 10$ A. The minimum (de)focusing length due to the (anti)proton beam is

$$F_{a,p} = \sqrt{\frac{\gamma_e\beta_e^2 m_e c^2\sqrt{2\pi}\sigma_{a,p}^2\sigma_s}{e^2 N_{a,p}(1+\beta_e)}}. \tag{2.11}$$

For electron lens with 10 kV 2 A electron beam one gets $F_e = 0.64$ cm, while $F_B = 0.066$ [cm]/B[T] and $F_{a,p} = 0.6$ [cm] if $N_{a,p} = 10^{11}$, $\sigma_{a,p} = 1$ mm, and $\sigma_s = 15$ cm (the rms bunch length). The beam envelope oscillations are stable if the focusing terms in the (2.8) are stronger than the defocusing terms, i.e., in the worst case of operation with antiprotons like in the Tevatron:

$$\frac{1}{F_B^2} \geq \frac{1}{F_e^2} + \frac{1}{F_a^2}. \tag{2.12}$$

For the parameters of the numerical example with 10 kV electron beam considered above, the required magnetic field is 0.15 T. Since the device uses the electron beam once over a passage, in principle one could consider having no magnetic field at all if the electron beam energy is high enough to have only minor electron beam disruption over the length of the (anti)proton bunch $F_{a,p} \gg \sigma_s$. For $\sigma_s = 15$ cm this yields an electron kinetic energy of $U_e \gg 12$ MeV. Taking into account the high average current needed, one can see that the electron beam power would exceed dozens of MW. This makes the use of relativistic electron beams very impractical for the beam-beam compensation.

As we will discuss below, a much stronger solenoid field of the order of 3–6 T is necessary to maintain stability of the antiproton beam and reduce x–y coupling caused by the electron beam distortions. In such a strong field, magnetic focusing dominates the electron dynamics. Each electron undergoes very fast Larmor oscillations with the frequency $\omega_L = eB/(\gamma_e mc)$ and spatial period of $\lambda_L = 2\pi v_e/\omega_L = \pi F_B \approx 2.1$ [mm]/B[T] ~ 0.5 mm. Their orbits can be represented as tiny (micron scale) Larmor circles moving along the magnetic field line.

The effect of a space-charge field E is that each of these circles starts to rotate slowly (drift) around the beam axis while staying at the same radius, i.e., the round beam remains round. The drift velocity in crossed electric and magnetic fields E and B is equal to

$$\vec{v} = c \frac{[\vec{E} \times \vec{B}]}{B^2}. \tag{2.13}$$

The radial space-charge electric field inside a constant current density electron beam with $j_e = J/\pi a_e^2$ is proportional to the radius $E = 2j_e r/\beta_e$. The angle θ_d of the drift rotation over the time interval t does not depend on the radius $\theta_d = v_d t/r = 2j_e ct/\beta_e B$. For example, the angle over the beam passage of $L = 2$ m in a $B = 4$ T field is about $\theta_d \approx 240°$. However, the electric field of a Gaussian (anti) proton beam is not linear, and the rotation angle θ_d is no longer independent of r, and electrons with larger r perform drift rotation on different (smaller) angles. The difference is negligible for our parameters—see the detailed studies in [24]. The magnetic forces due to electron and (anti)proton currents produce additional drifts similar to electric ones, but their contributions are β^2_e and β_e times smaller, respectively, and therefore negligible.

The required current density for a ~2 mm diameter beam is $j_e = J/\pi a_e^2 \approx 315 \beta_e/(1 + \beta_e)$ A/cm^2, or about 53 A/cm^2 for a 10 kV electron beam, $\beta_e = 0.2$. On the other hand, the oxide cathode lifetime deteriorates greatly if the current density exceeds 5–10 A/cm^2 (see [25] and references therein). To get within the cathode current density limit, one can use adiabatic magnetic compression in which the beam is born on the cathode with a larger radius a_c in a weak field B_c and transported to the region of stronger magnetic field B, with conservation of the adiabatic invariant $B_c a^2_c = Ba^2$. For the electron lens with cathode current density of about 2.1 A/cm^2 and $a_c = 5$ mm, the maximum "compression" ratio $R = B/B_c = a^2_c/a^2$ should be about 25, e.g., $B = 4$ T, $B_c = 0.16$ T. As we see in the following sections, such magnetic fields are quite feasible technically and were achieved in the Tevatron and RHIC electron lenses. Nonlinear head-on beam-beam compensation and the footprint compression also require precise control of the transverse electron charge distribution $\rho_e(r)$. This can be done by using near-cathode electrodes in the diode electron gun. If one applies a potential to these electrodes which is different from the cathode potential, then the distorted electric field distribution on the cathode surface will decrease (or increase) the electron emission from different radial areas of the cathode.

2.2.1.3   Effect of Ions

While passing through the vacuum chamber, the electron beam ionizes residual gas atoms and produces electrons and positively charged ions. Under certain conditions both these electrons and ions may accumulate in the electron beam. This could result in (a) change in the total charge density within the beam, i.e., changing the effectiveness of the beam-beam compensation, and (b) development of the so-called two-beam drift instability. Proper removal of the secondary particles is needed. Ionization of residual gas by electrons produces ions with the rate

$$\frac{dn_i}{dt} = \sigma_{ioniz} n_e v_e n_0, \tag{2.14}$$

where σ_{ioniz} is the ionization cross section, n_e is the electron beam density, and n_0 is the residual gas density. A useful quantity is the "neutralization time"

$$\tau_n = \frac{1}{\sigma_{ioniz} v_e n_0}, \tag{2.15}$$

the time in which the electron beam is fully neutralized if all produced ions remain in the beam. For $\beta_e = 0.2$ (10 kV) the total ionization cross section for N_2 is approximately $\sigma_{ioniz} \approx 2 \times 10^{-17}$ cm^2 [26]. At room temperature the residual gas density is $n_0 \approx 3.2 \times 10^{16} P$ [Torr] cm^{-3}. So, the neutralization time is approximately

$$\tau_n = \frac{2.5 \times 10^{-10} \text{ s}}{P[Torr]}, \tag{2.16}$$

For example, $\tau_n \approx 0.25$ s if $P = 10^{-9}$ Torr.

Drift instability is the main limitation on the electron beam current in the presence of ions. The origin of the phenomenon is the exponential amplification of a small initial separation of the electron and ion beams along the beam at certain frequencies. This amplification results in an instability if the amplification coefficient is larger than the feedback coefficient from the beam end to the beam beginning. Theoretical analysis of the instability is given in [27, 28] and agrees well with experimental investigations of the fully neutralized ($n_i = n_e$) electron beam [27]. The stability threshold current density is found to be about

$$j_e = \frac{v_e^2 B}{3.8 Lc}, \tag{2.17}$$

Taking this expression one can estimate the acceptable ion density for the electron compressor

$$\frac{n_i}{n_e} < \frac{v_e^2 B}{4Lcj_e} \approx 8\frac{B[T]\,\beta_e^2 a_e^2\,[mm]}{L[m]J_e[A]}, \tag{2.18}$$

If $\beta_e = 0.2$, $B = 4$ T, $a_e = 1$ mm, $J_e = 2$ A, $L = 2$ m, then $n_i/n_e < 0.13$. Therefore, though the ions should be cleaned out from the beam in order to avoid instability, the instability threshold value is rather relaxed. A more stringent requirement comes from the condition that the total charge distribution remain controllable within at least a few percent, i.e., $n_i/n_e < 0.01$. Originally, it was thought that special types of cleaning electrodes would be needed in the electron lenses—see, e.g., discussion and Figs. 9 and 10 in [3], but the operational experience with the Tevatron electron lenses showed that under conditions of the high beam vacuum in the collider, the ion accumulation posed little issue and the electrodes were rarely employed.

One should also mention several possible problems associated with the ionization electrons. These electrons are highly magnetized, so they can move only longitudinally and drift around the beam. Since both the electron gun and collector have negative potential with respect to the main vacuum chamber, the electrons could be trapped. However, there several mechanisms that force them to leave the system. First, they experience vertical gradient drift proportional to $[B \times \mathbf{grad}\ B]$ every time they pass through the bended section of the lens. Second, they are heated by the main electron beam until their energy increases enough to leave the potential well. The heating rate of this "electron wind" is $d(W/U_e)/dt \approx 4L_c e^2/mv_e a^2\,[1 + 2\ln(b/a)]$, where $L_c \sim 7$ is the Coulomb logarithm. Third, the center of the electron trajectory jumps by approximately a Larmor radius each time the electron is reflected from the potential wells of the electron gun or collector (if the Larmor circle step is larger than the length over which the potential changes). All that leads to a diffusive (or systematic if the axial symmetry is not perfect) loss of these electrons.

2.2.2   Side Effects on High Energy Beams

2.2.2.1   Electron Beam Distortions in Beam-Beam Compensation Setup

As shown in the previous section, collision with a round (anti)proton bunch in a strong magnetic field conserves axial symmetry and only minimally affects the radial size of the electron beam. Therefore, the electron beam space-charge forces are the same for antiprotons at the head and at the tail of the antiproton bunch. This is no longer true if the electron or (anti)proton beam is not round. Axial symmetry of the electron beam can be assured by using a round cathode in the electron gun and by making an appropriate choice of the magnetic field in the transport section of the setup. The (anti)proton beam roundness cannot be guaranteed *a priory*; moreover, for some applications, such as long-range beam-beam compensation, it is required to have two electron lenses at the machine locations with unequal vertical and horizontal beta functions β_x β_y, and, correspondingly, different vertical and horizontal (anti)proton sizes. Therefore, the electron beam cross section becomes a rotated ellipse as the tail of a non-round (anti)proton bunch passes through it, whereas the head of the bunch sees the original undisturbed round electron beam. Detailed numerical studies of the effect can be found in [24]. The electron beam distortions are of concern because (a) the distortion of the electron space-charge forces which might undermine the effectiveness of the beam-beam compensation, (b) in addition to the desired defocusing effect, electric fields of the elliptic electron beam produce $x-y$ coupling of vertical and horizontal betatron oscillations in the (anti)proton beam, and (c) a "head-tail" interaction appears in the antiproton bunch via higher order wake fields propagating in the electron beam. The electron beam distortions can be found analytically. We start with the continuity equation for the electron charge density ρ_e (x, y, s, t):

$$\frac{\partial \rho_e}{\partial t} + div\left(\rho_e \vec{v}\right) = 0, \tag{2.19}$$

where $v(x, y, s, t)$ is the velocity of electrons. Since the longitudinal component of the velocity is constant $v_s = \beta_e c$ and all longitudinal scales like the (anti)proton bunch length σ_s or electron beam length are much bigger than the transverse scale, one can neglect the term $d(\rho_e v_s)/ds$ in (2.19). We have separated the fast small amplitude Larmor motion and the slow large amplitude drift with velocity v_d in (2.13). The latter is the major source of the electron beam distortion, and in the following analysis we consider $v = v_d$. Assuming the unperturbed electron charge distribution to be axially symmetric ρ_e $(t = 0) = \rho_0(r)$, and that the maximum density distortion is small $\rho_e = \rho_0 + \delta\rho$, $\delta\rho \ll \rho_0$, then in the lowest order one gets from (2.19):

$$\frac{\partial \delta\rho_e}{\partial t} + \vec{v}_d \cdot grad\rho_0 + \rho_e div(\vec{v}_d) = 0. \tag{2.20}$$

The third term is equal to zero because div $v_d = 0$. The gradient in the second term can be written as grad $\rho_0 = 2r \, d\rho_0(r^2)/d(r^2)$ and thus, we obtain

$$\vec{v}_d \cdot grad\rho_0 = \frac{2c}{B^2} \frac{d\rho_0(r^2)}{d(r^2)} [\vec{E} \times \vec{B}] \cdot \vec{r}. \tag{2.21}$$

The electric field of the round electron beam does not contribute to the product above since it is proportional to r. The contribution due to electron beam space charge can be ignored as long as the electron charge density distortions are small with respect to $\rho_0(r)$. The major reason for the density change $\delta\rho$ is the (anti)proton beam space-charge force. The electric field of the elliptic Gaussian relativistic (anti) proton beam with rms sizes σ_x and $\sigma_y = R \, \sigma_x$ is given by

$$\vec{E} = -eN_{a,p}\lambda(s) \cdot grad, \tag{2.22}$$

where the linear density of antiprotons is normalized as $\int \lambda(s)ds = 1$, and the two-dimensional effective interaction potential $U(x, y)$ is [29, 30]:

$$U(x, y) = \int_0^\infty dq \frac{1 - e^{-\frac{x^2}{2\sigma_x^2(1+qR)} - \frac{y^2}{2\sigma_y^2(1+q/R)}}}{\sqrt{(1 + qR)(1 + q/R)}}. \tag{2.23}$$

Therefore, after some mathematics we get

$$\delta\rho\left(x, y, t = \frac{s}{(1+\beta_e)c}\right) = \left(\int_{-\infty}^s \lambda(s')ds'\right) \frac{2eN_{a,p}}{B} \frac{d\rho_0(r^2)}{d(r^2)}$$

$$\times \frac{xyI(x, y)\left(\sigma_x^2 - \sigma_y^2\right)}{\sigma_x^2 \sigma_y^2}, \tag{2.24}$$

where now s is the coordinate inside the (anti)proton bunch ($s = -\infty$ for the bunch head) and

$$I(x, y) = \int_0^\infty dq \frac{1 - e^{-\frac{x^2}{2\sigma_x^2(1+qR)} - \frac{y^2}{2\sigma_y^2(1+q/R)}}}{(1 + qR)^{3/2}(1 + q/R)^{3/2}}, \tag{2.25}$$

e.g., $I(0, 0) = 2R/(1 + R)^2$. The major features of the distortion are (a) it is absent in the case of a round antiproton beam when $\sigma_y = \sigma_x$, (b) it performs two variations over azimuth as $\delta\rho$ scales as $xy \sim \sin(2\theta)$, (c) it diminishes as the solenoid field B increases, or as the (anti)proton intensity $N_{a,p}$ decreases, and (d) most of the distortion takes place at the radial edge of the electron beam and since grad $d\rho_0(r^2)/d(r^2) \sim \rho_0^{max}/a^2$, a wider electron beam receives smaller density distortions during the interaction; there is also no distortion inside a round constant density electron beam. Finally, the scaling of the maximum distortion strength is

$$\frac{\delta\rho^{max}}{\rho_0^{max}} \cong \frac{0.2eN_{a,p}}{a_e^2 B} \approx \frac{0.1N_{a,p}[10^{11}]}{a_e^2[mm]B[T]}, \tag{2.26}$$

where the value of 0.2 comes from the geometrical factor $\sim xy\, I(x, y)$. For example, the distortion is about 1.5 % for a 1 mm radius electron beam in a $B = 4$ T solenoid field. Note that as soon as the elliptic distortion appears it starts a drift rotation in the crossed fields of the electron space charge and the solenoid field. It is important that during the passage of the (anti)proton bunch—typically, few ns—the rotation is small. For example, for $B = 2$ T the angle is about $\theta_d \approx 4j_e\sigma_z/\beta_e B \sim 0.1$ rad $\ll 1$.

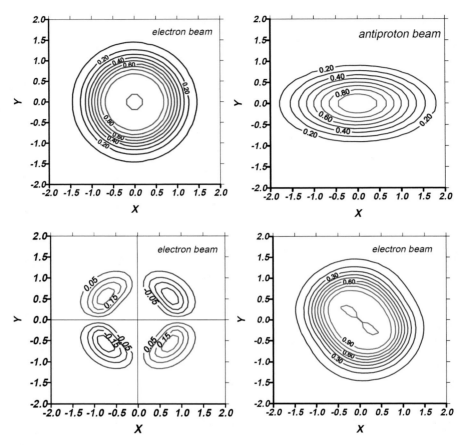

Fig. 2.2 Contour plots of original electron density (*top left*), antiproton density (*top right*), change of the electron density due to interaction with antiproton space charge (*bottom left*), and resulting electron density (*bottom right*). x and y coordinates are given in units of σ_x [24]

Thus, ignoring the second factor in (2.20) is justified. Figure 2.2 shows the electron beam distortion calculated analytically with (2.24). The top left-hand plot in Fig. 2.2 shows contour lines of constant density for the electron beam with a particular initial distribution of

$$\rho_0(r) = \frac{1}{1 + (r/a_e)^6}, \quad a_e = \sigma_x = 0.6\,\text{mm}. \tag{2.27}$$

Constant density lines for the Gaussian distribution in the (anti)proton beam with $\sigma_x = 0.61$ mm and $\sigma_y = 0.31$ mm are presented in the top right-hand plot. The lower left-hand corner of the figure shows the change of the electron charge density $\delta\rho(x,y)$ after passage through an (anti)proton bunch with $N_{a,p} = 6 \times 10^{10}$ in a magnetic field $B = 0.4$ T. With such a small solenoid field the distortion is very large $\delta\rho$ ~0.25 and the resulting electron beam shape $\rho = \rho_0 + \delta\rho$ is clearly a rotated ellipse as shown in the lower right-hand plot. In this case, the space-charge fields are very different for the (anti)protons in the head and in the tail of the bunch. The solenoid field in the electron lens setup will be about ten times stronger $B \approx 4$ T and consequently some ten times smaller electron beam distortions are expected.

For some applications such as long-range beam-beam tuneshift compensation, the electron beam has to be 2–3 times wider than the (anti)proton beam to assure adequate electron space charge force linearity in the electron lens setup. According to (2.26), that helps to reduce $\delta\rho$ by another factor of 4–9 [24].

2.2.2.2   Coupling Due to Distorted Electron Beam

Electric and magnetic fields of the elliptic electron beam lead to $x-y$ coupling of vertical and horizontal betatron oscillations in the antiproton beam. Since originally the electron beam is round, the head of the (anti)proton bunch experiences no coupling force. But, according to the above analysis, the electron density distortion grows as $\int \lambda(s')ds'$—see (2.24)—and the coupling grows proportionally. Particles in the head and in the tail of the high energy bunch change their positions while performing synchrotron oscillations, thus an average coupling effect is about half of the maximum coupling spread. The average coupling can be corrected in the machine, but there are no available tools to compensate the spread in the coupling. Therefore the coupling spread has to be small enough in order not to affect the antiproton beam dynamics. For example, satisfactory operation of the Tevatron collider required global coupling parameter $|\kappa|$ to be corrected to better than few 0.001. Detailed analysis of the coupling effect due to the electron beam [3] gives an upper limit of the effect

$$| \kappa | < \frac{dQ_{EL} e N_{a,p}}{\sigma_x^2 B} S^{max}, \tag{2.28}$$

where dQ_{EL} is the tuneshift due to electron lens (1.17) and maximum numerical factor S^{max} of the order of 0.1 depends on the ratio of the electron and (anti)proton beam sizes. For typical parameters of the Tevatron $\sigma_x = 0.61$ mm, $N_a = 6 \times 10^{10}$, $B = 4$ T and $dQ_{EL} \sim 0.01$ the maximum coupling spread was found to be about $|\kappa| \approx 2 \times 10^{-4}$ for a narrow electron beam with $a_e = \sigma_x$, and $|\kappa| \approx 3.5 \times 10^{-5}$ for a wider electron beam with $a_e = 2.5 \sigma_x$. Both of these values rather negligible to affect the collider operation. Note that two 2 m long 4 T solenoids of the Tevatron electron lenses induce a static coupling of about $|\kappa| \approx 0.001$ that was easily corrected in operations.

2.2.2.3   Head-Tail Effect Due to Electron Beam

This section is devoted to the stability of the high energy (anti)proton beam interacting with low energy electron beam in the electron lens setup. Electron space-charge forces cause transverse head-tail coupling within the (anti)proton bunch which may lead to a transverse mode coupling instability (TMCI). In this Section, we mainly follow the article [31] by A.Burov, V.Danioopv and V.Shiltsev where a detailed theory, analytical studies, and numerical simulations of the effect can be found.

Low energy electrons can create significant transverse impedance comparable with the intrinsic impedance of the collider ring, and this can result in a collective instability of the (anti)proton bunch. In the most known case of so called electron-cloud instability [32], the secondary electrons accumulated in high energy beams lead to instabilities and limit performance of high current proton accelerators, including such colliders as RHIC and LHC. The situation with the electron lens is quite different as the electron beam is born on an electron gun cathode, transported through the interaction region, and absorbed in the collector. Therefore, each portion of electrons passes through the (anti)proton beam only once and there is no multi-turn memory in the system of two interacting beams and only short distance transverse wake fields are of concern. The phenomenon is as follows: If the centroid of the antiproton bunch head collides off the electron beam center, then the electron-proton attraction (repulsion in the case of electron-antiproton interaction) causes the electron motion. As a result, the electron beam has a displacement when it interacts with the tail of the bunch. The impact of the electron beam on the following (anti)protons depends on the transverse coordinate of the preceding (anti) protons. Such a head-tail interaction leads to the TMCI. This effect is similar to the "strong head-tail" interaction via vacuum chamber impedance first observed a long time ago in electron storage rings [33]. The TMCI in the electron rings limits the maximum single bunch current. In our case, the source of the coupling is the electron space charge which is the basic mechanism for the beam-beam compensation and other electron lens applications and thus, cannot be avoided. The way to counteract the instability is to increase the electron beam rigidity making its motion during the collision smaller. Naturally it can be done using a strong longitudinal magnetic field in the interaction region. (We assume the collider ring chromaticity can be made small enough—as it typically is—so that the "weak head-tail" [33] instability is non-existent.)

Threshold longitudinal magnetic field necessary to avoid the instability and the dependence of the threshold on electron and (anti)proton beam parameters are estimated below. Three types of analysis—simplified theoretical wake-field model, multi-mode analysis and direct numerical simulations—all end up with similar results.

Theoretical model of direct and skew wakes: Conventionally, the analysis of relativistic beam stability relies on the wake function concept; see, e.g., [33]. Electromagnetic fields excited in an accelerator vacuum pipe vary over transverse distances of about the pipe aperture b, which is usually much larger than the beam radius a. This allows an expansion of the perturbation on dipole, quadrupole, and higher order terms over a small parameter (a/b). The situation is different for the case under study. The electron beam space-charge fields excited by (anti)protons have about the same transverse extent as the (anti)proton beam, and this complicates the analysis. However, the interaction can be described by the conventional approach for a specific case when both the (anti)proton bunch and the electron beam are homogeneous and bounded by the same radius $a = a_p = a_e$. In this case, electromagnetic wake fields have a simple radial structure and can be easily calculated. To find the dipole wake function, let us consider a thin antiproton slice with a charge q and offset Δx traveling through the electron beam. After interacting with the slice, electrons acquire a transverse velocity given by

$$v_{xe} = \frac{2eq\Delta x}{a^2(1+\beta_e)\gamma_e mc}. \tag{2.29}$$

Such a kick causes transverse Larmor oscillations in the longitudinal magnetic field B and, after a time interval t, the electron transverse offsets are

$$x_e = \frac{v_{xe}}{\omega_L}\sin(\omega_L t), \quad y_e = \frac{v_{xe}}{\omega_L}[1 - \cos(\omega_L t)], \tag{2.30}$$

where $\omega_L = eB = \gamma_e mc$ stands for the Larmor frequency. One can see that an originally horizontal displacement Δx results in both horizontal and vertical displacements. Taking into account the possibility of a vertical offset y, we conclude that antiprotons at a distance s behind the slice will experience momentum changes equal to

$$\Delta p_x(s) = -\frac{eq}{c}[W_d(s)\Delta x - W_s(s)\Delta y], \quad \Delta p_y(s) = -\frac{eq}{c}[W_s(s)\Delta y + W_d(s)\Delta y], \tag{2.31}$$

where we introduce the direct wake function $W_d(s)$ and the skew $W_s(s)$ wake function

$$W_d(s) = W\sin(ks), \quad W_s(s) = W[1 - \cos(ks)], \tag{2.32}$$

$W_d(s) = 0$, if $s < 0$, and

$$W = \frac{4\pi n_e L_e}{(1+\beta_e)a^2(B/e)}, \quad k = \frac{\omega_L}{(1+\beta_e)c}, \tag{2.33}$$

where $n_e = J_e/(e\pi a^2 v_e)$.

Depending on the parameters, one or another of the two wake functions (2.32) can have a dominant influence on the (anti)proton beam stability. The direct wake effects are suppressed if there are many Larmor oscillations periods over the antiproton bunch length ss, while the skew force impact decreases with increasing $x-y$ detuning. In the case of the Tevatron operating near the coupling resonance $\Delta Q = |Q_x - Q_y| < 0.01$, consideration of the coupling of only the two closest modes $Q_x + mQ_s$ and $Q_y + nQ_s$ (m, n are integer) gives the following expression for the threshold magnetic field [31]:

$$B_{thr} \approx 1.3 \frac{eN_{a,\,p}\sqrt{dQ_x^e dQ_y^e}}{a^2 \sqrt{\Delta Q \cdot Q_s}}, \tag{2.34}$$

where the horizontal and vertical tuneshifts $dQ_{x,y}^e$ induced by the electron lens are given by (1.17), and Q_s is the collider synchrotron tune. Let us take for numerical example, $dQ_x^e = dQ_y^e = 0.01$, $N_{a,p} = 6 \times 10^{10}$, $Q_s = 0.001$, $\Delta Q = |Q_x - Q_y| = 0.01$, $a = 1$ mm, then the threshold solenoid magnetic field has to be more than $B_{thr} = 1.2$ T.

Multimode analysis: The two mode coupling model allows us to derive analytical formulas for the TMCI threshold by taking into account only a constant skew component of the wake force due to the electron beam and just two coupling modes. A more general numerical algorithm for calculating the mode coupling is developed in [34] and it avoids such simplifications and considers many modes and a general wake form and most importantly, it deals with non-averaged motion. In that analysis, the antiproton bunch is divided into several radial and azimuthal parts in synchrotron phase space and consequently, a series of synchrobetatron modes can be seen. The wake force kick changes the backward particles' angles. The rest of the accelerator is presented by a linear transformation matrix (rotation in phase space). Eigenvalues (eigentunes) of the resulting transformation matrix can be calculated numerically. The complexity of the calculations is squared as the number of modes, so for calculations with MATHCAD software one has to limit the number.

The results presented in Figs. 2.3 and 2.4 are obtained with the high energy antiproton bunch divided into 1 radial (i.e., the same synchrotron oscillations amplitude for all particles) and 7 azimuthal parts for both vertical and horizontal degrees of freedom. Thus, it is possible to see the behavior of the first 1 radial and 7 azimuthal synchrobetatron modes in horizontal and vertical motion taking into account their coupling. Complete expressions for the linearized direct and skew transverse wake functions, (2.32) are used. Numerical parameters used in these calculations are $N_a = 6 \times 10^{10}$, the rms size of the round Gaussian antiproton beam is $\sigma_p = 1$ mm, and the longitudinal magnetic field is equal to 1 T. Figure 2.3 shows the calculated eigentunes versus the linear betatron tune shift dQ^e induced by the interaction with electron beam. The fractional part of the machine's betatron tune for the horizontal motion is $Q_x = 0.556$ and for the vertical $Q_y = 0.555$, and the synchrotron tune Q_s is 0.001. Therefore, the betatron tune difference is comparable with the synchrotron tune. In the absence of the electron current, $dQ^e = 0$ and the eigenfrequencies of the azimuthal modes are equal to $Q_{x,y} + kQ_s$, where the integer k has seven values in the range of $-3,\ldots,3$ and represents the number of modulation periods in synchrotron phase space. Some of the modes coalesce with increasing

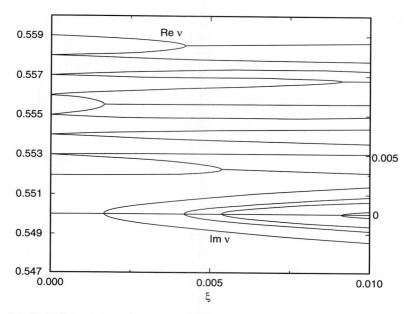

Fig. 2.3 Eigen frequencies (*tunes*) of the antiproton bunch oscillation modes vs. the antiproton betatron tune shift due to the electron beam (*horizontal axis*). The vertical scale on the left is for the fractional part of the tunes Ren (*upper series of lines*), and the *right-side scale* is for the imaginary part of the tunes Imn (*lower series of lines*) [31]

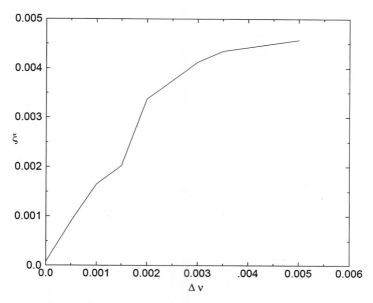

Fig. 2.4 Threshold antiproton tune shift dQ^e due to the electron beam (*vertical axis*) vs. the difference of antiproton horizontal and vertical tunes $\Delta v = |Q_x - Q_y|$. $B = 1$ T, $Q_s = 0.001$, $N_a = 6 \times 10^{10}$ [3]

dQ^e, and the real parts of their tunes ReQ (see upper series of curves in Fig. 2.3) become equal, while the imaginary parts ImQ bifurcate into one negative and one positive branch. The latter evidently means instability in the motion. In our case, the first merging of modes takes place at $dQ^e \approx 0.0017$, the next merging of higher modes occurs at $dQ^e \approx 0.0045$, etc.

Figure 2.4 shows the tune shift threshold dQ^e for the first coupling modes versus the tune split $\Delta\nu = |Q_x - Q_y|$ for a vertical tune equal to 0.555. The threshold grows linearly until $\Delta\nu \sim 2$–2.5 and then it is approximately proportional to $\Delta\nu^{1/2}$—in good agreement with the two-mode model. Note that a completely adequate consideration of the fast oscillating parts of the wakes would require many more modes $\sim k\sigma_s \approx 30$–100 to be taken into account [3].

Numerical simulations of the TMCI due to electron beam: Three-dimensional simulations of the effects have been done also with a numerical code [31]. The round Gaussian antiproton beam ($\sigma_x = \sigma_y = \sigma_a$) is presented as a number of macroparticles (typically in the range from $M = 128$ to maximum 2048). The particles have equal charges $e\Delta N_a = eN_a/M$. Both direct and skew wakes are taken into account in this numerical simulation. The simulation reveals that although the antiproton bunch motion is essentially two-dimensional (since the wake is 2D), the instability starts in that plane where the original lattice tune is closer to half integer $Q = 1/2$, e.g., in the horizontal plane for the Tevatron ring. Figure 2.5 shows results of the numerical simulations giving the threshold strength

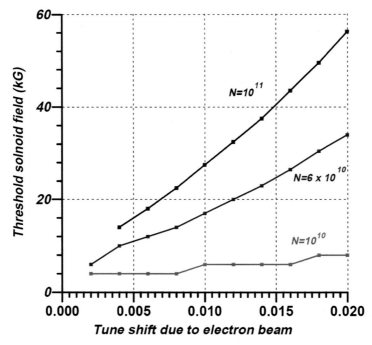

Fig. 2.5 Threshold solenoid field B_{thr} vs. tune shift due to electrons dQ^e at different bunch populations $N_a = 1,6,10 \times 10^{10}$. Focusing lattice tunes $Q_x = 0.585$, $Q_y = 0.575$, synchrotron tune $Q_s = 0.0012$, no betatron tune spread in the beam, and the rms size of antiproton beam $\sigma_a = 0.7$ mm [31]

of the solenoidal magnetic field B_{thr} vs the electron beam tuneshift parameter dQ^e for antiproton bunch populations equal to $N_a = (1,6,10) \times 10^{10}$—lower, middle, and upper curves respectively. The threshold is defined as the value of B which results in more than a tenfold increase of the initial centroid betatron amplitude over the first 10,000 turns. One can see that the field is approximately proportional to both dQ^e and N_a in accordance with the theoretical prediction (2.34). It was found that the dependence of the threshold on the synchrotron tune agrees well with theory, also, i.e., $B_{thr} \sim Q_s^{1/2}$.

In order to evaluate the importance of the oscillation part of the wakes in (2.32) similar simulations are performed without the constant part of the skew wake, i.e., with $W_d(s) = W \sin(ks)$ and $W_s(s) = W \cos(ks)$. It is found that in this case, about five times smaller solenoidal field is needed to assure beam stability. This confirms the decisive role of the constant part of the skew wake that is a basic assumption of the two-mode coupling model.

The TMCI threshold sensitively depends on the operation point (Q_x, Q_y). Figure 2.6 presents the results of scanning the horizontal tune Q_x from 0.52 to 0.63 while the vertical tune is held constant at $Q_y = 0.575$. In close vicinity to the coupling resonance $|Q_x - Q_y| < 15\ Q_s$, the threshold magnetic field depends on Q_s approximately as $\sim 1/|Q_x - Q_y|^\kappa$, where $2/5 < \kappa < 1/2$. Away from the resonance, the best fit power is $\kappa \approx 1/5$. The threshold also goes up near the half-integer resonance $Q_x \rightarrow 0.5$. In order to compare simulations with the two-mode model, one can fit B_{thr} in a form similar to (2.34)

$$B_{thr} \approx 0.95 \frac{eN_a dQ^e}{\sigma_a^2 \sqrt{|Q_x - Q_y| \cdot Q_s}}, \tag{2.35}$$

see also the dashed line in Fig. 2.6.

The differences in numerical factors between (2.35) and (2.34) are due to (a) the kick due to a Gaussian beam differs from a round beam kick (2.29) and σ_a is used instead of a, (b) the oscillating parts of the wake forces are taken into account in the simulations in contrast to the two mode model, and (c) more than two modes play a role in the computer tracking because of the large number of macroparticles. Nevertheless, there is very good quantitative agreement with the results of the multimode analysis presented in Fig. 2.4.

Neither the two-mode theory nor the multimode analysis deal with the tune spread in the antiproton bunch, which tends to suppress the instability due to Landau damping [36]. The TMCI can be additionally suppressed if the electron beam has a radius larger than the antiproton beam radius, $a > \sigma_{a,p}$. It is shown in [31] that the skew wake function scales with the electron beam radius as $W_s \sim n_e/a^2$; i.e., a two times wider electron beam will lead to a four times smaller required magnetic field for the same dQ^e. The oscillating direct wake function $W_d(s)$ does not depend on the electron beam radius when the electron density is fixed.

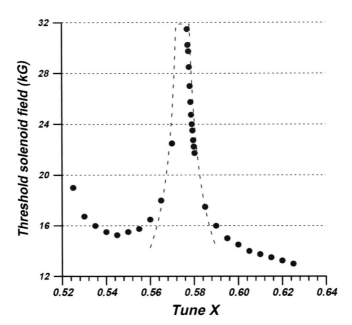

Fig. 2.6 Threshold magnetic field vs horizontal tune Q_x. Dashed line corresponds to $B_{thr} \sim 1/|Q_x - Q_y|^{1/2}$, $Q_y = 0.575$, $Q_s = 0.001$, $dQ^e = -0.01$, $N_a = 6 \times 10^{10}$, $\sigma_a = 0.7$ mm [31]

2.2.2.4 Effect on Another Beam in Single Aperture Colliders

Proton-antiproton colliders offer a unique opportunity to circulate protons and antiprotons in the same magnetic system and in the same beam pipe, as it was in the Tevatron collider. Therefore, the operation of the electron lens on one of the beams, say, on antiprotons, would potentially affect the proton beam as well. Let us assume that direction of the electrons' propagation is opposite to the antiprotons velocity (i.e., they collide). The proton beam moves in the opposite direction in the same vacuum chamber and also may effectively interact with the electron current. If the proton and antiproton beam orbits are not separated, an additional positive tune shift for protons is

$$dQ_p^e \approx \frac{\beta_z N_e r_p (1 - \beta_e)}{2\pi a_e^2 \gamma_p} = -dQ_a^e \frac{1 - \beta_e}{1 + \beta_e}, \qquad (2.36)$$

where $\beta_z = \beta_{x,y}$ is vertical or horizontal beta-function at the electron lens location, $N_e = J_e L / \beta_e c$ is the total number of electrons on the orbit protons and antiprotons going through—compare with (1.17). If $\beta_e \ll 1$, dQ_p^e does not differ too much from $dQ_a^e \sim 0.01$ and might be too large to tolerate. But, for very similar concerns of beam-beam interaction, the proton and antiproton orbits are well separated in all but two interaction points. Correspondingly, with the electron lenses installed away from main IPs, the impact on the proton beam due to electron beam, assuming the latter centered on antiprotons, is minimized, too:

$$dQ_p^e \approx -2 \frac{dQ_a^e}{(d/a_e)^2} \frac{1-\beta_e}{1+\beta_e},\tag{2.37}$$

where d is the separation of the proton and electron beams. For example, a 7 mm orbit separation from 1 mm diameter electron beam makes dQ_p^e about 25 times smaller than dQ_a^e. In the Tevatron operation that was sufficient to avoid detectable impact on protons (or, vice versa, on antiprtons if the electron lens was centered on proton beam), but in more difficult cases one may consider additional suppression by pulsing the electron current only on one type of species without affecting the other.

2.2.2.5 Electron Current Fluctuations

Fluctuations of the electron current from turn to turn cause time variable quadrupole kicks which lead to a transverse emittance growth of the (anti)proton bunches. The current in the electron lens setup has to be modulated rather fast although periodically, and thus, the issue of how stable the current is at a one-turn scale may be of importance.

The emittance growth rate due to fluctuations of a gradient δG of a lens with length L is given by [35]

$$\frac{d\varepsilon_z}{dt} = f_0^2 \frac{\varepsilon_z}{16} \left(\frac{eL\beta_z}{Pc}\right)^2 \sum_{n=-\infty}^{\infty} S_{\delta G}(f_0 \mid 2Q_z - n \mid),\tag{2.38}$$

where z stands for either x or y, f_0 is the revolution frequency, β_z is beta function at the lens location, P is the antiproton momentum, Q_z is the machine tune, and $S_{\delta G}(f)$ is the power spectral density (PSD) of the gradient fluctuations. One can see that only some particular frequencies contribute to the emittance growth, and the lowest of them is $(2Q_z-1)f_0$, or about 7 kHz for the Tevatron. If one assumes that the electron current ripple is "white noise" with a constant PSD $S_{\delta G}$, then the rms value of the ripple δG relates to the PSD as

$$\delta G^2 = \frac{1}{2} f_0 S_{\delta G}.\tag{2.39}$$

The effective gradient G of the electron lens relates to the tune shift as $dQ_z^e = (\beta_z/4\pi)(eGL_e/Pc)$ and, therefore, scales linearly with the electron, thus, combining (2.38) and (2.39) one gets:

$$\frac{d\varepsilon_z}{dt} = 2\pi^2 f_0 \varepsilon_z \left(dQ_z^e\right)^2 \left(\frac{\delta J_e}{J_e}\right)^2, \tag{2.40}$$

Where (dJ_e/J_e) is the rms value of relative current fluctuation. From (2.40) one gets the emittance evolution equation

$$\varepsilon_z = \varepsilon_{0z} \exp\left(t/\tau_z^e\right), \tag{2.41}$$

where the characteristic growth time is equal to

$$\tau_z^e = \frac{1}{4\pi^2 f_0 \left(dQ_z^e\right)^2 \left(\frac{\delta J_e}{J_e}\right)^2}. \tag{2.42}$$

For the parameters of the Tevatron of $f_0 = 48$ kHz and $dQ^e = 0.01$, one needs $(\delta J_e/J_e) < 0.5 \times 10^{-3}$ to have acceptable emittance growth time of about 10 h.

2.2.2.6 Transverse Electron Beam Motion

Transverse motion of the electron beam may also cause direct (anti)proton emittance growth. Indeed, if the electron beam displacement is equal to δX, then the dipole kick experienced by antiprotons is $\delta\theta = \delta X/F$, where $F = Pc/(eGL_e)$ is the focal length of the electron lens. Coherent (anti)proton betatron oscillations begin and after some decoherence time they result in (anti)proton emittance growth. The normalized emittance grows *linearly* in time and its growth rate is equal to [37]

$$\frac{d\varepsilon_z}{dt} = \frac{\gamma_p f_0^2}{4} \sum_{sources} \frac{\beta_z}{F^2} \sum_{n=-\infty}^{\infty} S_{\delta X}(f_0 \mid Q_z - n \mid). \tag{2.43}$$

Note that the lowest frequency of interest $|Q_z - n|f_0$ is the betatron frequency, that is about 20 kHz for the Tevatron. Using the same transformations as in the section above, one gets for the electron lens induced emittance growth:

$$\frac{d\varepsilon_z}{dt} = 8\pi^2 \gamma_p f_0 \cdot \delta X^2 \cdot \frac{(dQ_z^e)^2}{\beta_z}, \tag{2.44}$$

where δX now stands for the rms electron beam vibration amplitude. Let us constrain the emittance growth rate to be less than 0.25 π mm mrad/h, then for the electron lens tuneshift parameter $dQ^e = 0.01$ one gets the requirement on the rms electron beam turn-to-turn position stability $\delta X < 0.14$ μm. The tolerable amplitudes are several orders of magnitude larger than vibrations of the Tevatron quadrupoles at high frequencies [38].

If the electron beam and the (anti)proton beam are not properly aligned with respect to each other and they collide off center with displacement equal to ΔX, then the electron current ripple at the betatron frequencies causes dipole kicks on antiprotons and can also lead to transverse emittance growth. The tolerance can be easily estimated as

$$\frac{\Delta J}{J} \Delta X \approx \delta X. \qquad (2.45)$$

The tolerance depends on the straightness of the electron beam in the interaction region, which is determined by the solenoid field quality. Using $\Delta X \sim 0.15$ mm, one calculates the rms current ripple tolerance as $(\delta J_e / J_e) < 1.1 \times 10^{-3}$—that is a somewhat loose requirement in comparison with the quadrupole kick effect considered above.

2.2.2.7 Solenoid Field Quality

A strong solenoid magnetic field B of the order of several Tesla in the straight section of the electron lens assures that the electrons perform very small but fast Larmor oscillations around the magnetic field lines. Therefore, deviations of the magnetic field vector B from a straight line will cause off-center collisions of the antiproton and electron beams. In the case of the electron lens for nonlinear beam-beam compensation may cause unwanted nonlinear components of the electromagnetic forces. To avoid the effect, one needs to have the field lines not deviate from the straight antiproton orbit more than some part of the transverse (anti)proton beam size which is of the order of 1 mm in the Tevatron arcs. If one requires a solenoid field straightness equivalent to $\Delta X \sim 0.2$ mm, then a transverse field component has to be less than

$$\frac{\Delta B_\perp}{B} \sim \frac{\Delta X}{L} \sim \frac{0.2\,\mathrm{mm}}{2\,\mathrm{m}} = 10^{-4}. \qquad (2.46)$$

This is comparable with the requirement on the B-field quality in the electron cooling devices.

2.3 Practical Implementation of Electron Lenses in the Tevatron and RHIC

In this section we consider in detail main technologies and practices implemented in the operating electron lenses built for the Tevatron proton-antiproton collider and for RHIC. The Tevatron electron lenses (TELs) were used for long-range and head-on beam-beam compensation as well as for longitudinal and transverse beam

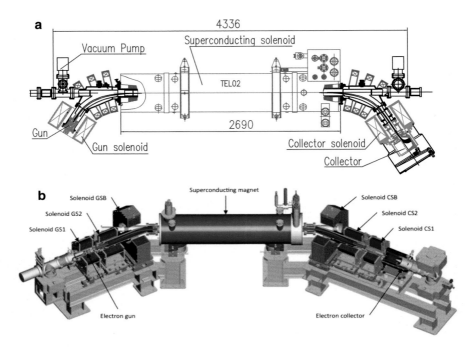

Fig. 2.7 (**a**) General layout of the TEL-2 installed at the Tevatron A11 location, *top view* [39]; (**b**) RHIC electron lens [40]. In both cases, the electrons move from *left* to *right* and interact inside the superconducting solenoid with the antiprotons (Tevatron) or protons (RHIC), which move in the opposite direction, or with protons in the Tevatron, which move in the same direction

collimation. Two lenses are built to compensate non-linear head-on beam-beam effects in RHIC. The Tevatron and RHIC systems have many similarities—see Fig. 2.7. Major relevant parameters of the colliders and the lenses are given in Table 2.2. This description of the electron lenses's technical systems mostly follows [39–41].

Two Tevatron Electron Lenses (TEL-1 and TEL-2) were built and installed in two different locations of the Tevatron ring, F48 and A11, in 2001 and 2004, respectively. Figure 2.7a depicts the general layout of the TEL-2. The electron beam is generated by a thermionic gun immersed in a solenoidal magnetic field. Strongly magnetized electrons are accelerated to a kinetic energy of 5–10 kV and follow the magnetic field lines into the main superconducting solenoid where the interaction with the high energy proton/antiproton bunches occurs. While the high energy particles continue on (along) the Tevatron orbit, the low energy electrons exiting the main solenoid are guided into the collector and are not being recirculated (reused). The lenses were used in three regimes of operation—(a) for compensation of beam-beam effects [42, 43]; (b) for removal of uncaptured particles from the abort gaps between the bunch trains [44], and (c) for transverse beam collimation by hollow electron beams [12]. Three conditions were found to be

Table 2.2 Main parameters of the Tevatron collider and RHIC, and their electron lenses (ELs)

Parameter	Symbol	Value		Unit
Electron Lenses:		*TEL*	*RHIC*	
e-beam energy (oper./max)	U_e	5/10	9.5	kV
Peak e-current (oper./max)	J_e	0.6/3	0.9	A
Magnetic field in main solenoid	B_m	3.1/6.5	5.0	T
Magnetic field in gun solenoid	B_g	0.29	0.3	T
e-beam radius in main solenoid	a_e	2.3	0.3 (rms)	mm
Cathode radius	a_c	7.5	7.5	mm
e-pulse repetition period	T_0	21	13	μs
e-pulse width, "0-to-0"	T_e	0.6	~0.4	μs
Interaction length	L_e	2.0	2.1	m
Maximum tuneshift by EL	$dQ^e_{a,p}$	0.009	+0.012	
Collider:		*Tevatron*	*RHIC*	
Circumference	C	6.28	3.83	km
Proton(*p*)/antiproton(*a*) energy	E	980	250	GeV
p- bunch intensity	N_p	270	250	10^9
a-bunch intensity (max.)	N_a	50–100	–	10^9
Number of bunches	N_B	36	111	
Bunch spacing	T_b	396	108	ns
p-emittance (normalized, rms)	ε_p	≈2.8	≈2.5	μm
a-emittance (normalized, rms)	ε_a	≈1.4	n/a	μm
Max. initial luminosity/10^{32}	L_0	4.3	2.5	10^{32} cm^{-2}s^{-1}
Beta functions at 1st EL (A11 TEL)	$\beta_{y,x}$	150/68	10/10	m
Beta functions at 2nd EL (F48 TEL)	$\beta_{y,x}$	29/104	10/10	m
p-head-on tuneshift (per IP)	ξ^p	0.010	0.012	
a-head-on tuneshift (per IP)	ξ^a	0.014	n/a	
Number of IPs	N_{IP}	2	2	
p-long-range tuneshift (max.)	ΔQ^p	0.003	n/a	
a-long-range tuneshift (max.)	ΔQ^a	0.006	n/a	

crucial for successful compensation of beam-beam effects by the electrons lenses: (a) the electron beam must be transversely centered on the proton (antiproton) bunches, within 0.2–0.5 mm, over the entire interaction length of about 2 m; (b) fluctuations in the electron current need to be less than one percent, and the timing jitter less than one nanosecond, in order to minimize emittance growth of the high-energy beams; and (c) the transverse profile of the electron current density should have a specific shape, depending of the application, e.g., a distribution featuring a flat-top and smooth edges is needed for compensation of the long-range beam-beam effects, hollow beam is needed for transverse collimation.

The head-on beam-beam interaction is the dominant luminosity limiting effect in polarized proton operation in RHIC. To mitigate this effect two electron lenses were installed in each of the two RHIC rings (marked as Blue and Yellow) in 2013–2014. The lenses compensate for only one of the two head-on collisions in

RHIC and allow significant increase in the proton beam intensity (see more details in Chap.3). Together with the RHIC polarized proton source and focusing optics upgrades, the electron lenses approximately doubled the machine luminosity during the 2015 *p-p* operations run. The Blue and the Yellow electron lenses installed side by side in one of the collider interaction regions, IR10, in a section common to both beams. Such arrangement allows local compensation of the main solenoid effect on both linear coupling and spin orientation by having the two main solenoids with opposing field orientations.

2.3.1   Magnetic and Cryogenic Systems

The main requirements for the magnetic system of an electron lens are formulated in Sect. 2.2 above. Besides being used for the transport of electrons from the cathode to the collector, the magnetic system must be capable of changing—by adiabatic magnetic compression—the electron beam size in the interaction region, and allow precise positioning of the electron beam on the high-energy beam of choice. Let us consider as an example the Tevatron electron lens magnetic system. The three solenoids in the TEL are oriented as shown in Fig. 2.7. The gun solenoid sits in the lower-left corner at the angle to the long Tevatron beam pipe, the main superconducting (SC) solenoid surrounds the beam pipe, and the collector solenoid resides in the lower right. The geometrical center of the Tevatron vacuum chamber is precisely aligned with the magnetic axis (center) of the main solenoid. The electrons, originating from the electron gun, follow the magnetic field lines, bent in the horizontal plane. The solenoids were manufactured at the Institute of High Energy Physics in Protvino, Russia and tested at Fermilab. Technical details on the magnet construction and magnetic field simulations can be found in [45, 46].

2.3.1.1   Main SC and Conventional Solenoids

The transverse cross section of the main TEL solenoid is shown in Fig. 2.8. It is capable of reaching a maximum field of 6.5 T at 1780 A and liquid Helium temperature of 4.6–5.3 K. The main solenoid does not contain a closed current loop; when energized, the current flows out of its current leads and through external power supplies. The main solenoid uses NbTi wire intertwined with copper wire (Cu/ NbTi ratio of 1.38), rated for 550 A at a temperature of 4.2 K; the wire itself measures 1.44 mm by 4.64 mm cross-section. The cable is wrapped with a polyimide film of 0.03 mm thickness with a 1/3 overlap. The main SC coil is wound on a stainless steel tube of 151.4 mm diameter and 4 mm thickness. The frame insulation is three layers of 0.1 mm thick polyimide film. A 4.85 cm-thick, low-carbon steel shield wraps around the coils, which enhances the field strength, keeps the field lines compressed near the solenoid's ends, improves the homogeneity throughout the interaction region, and reduces stray fields. While the solenoid

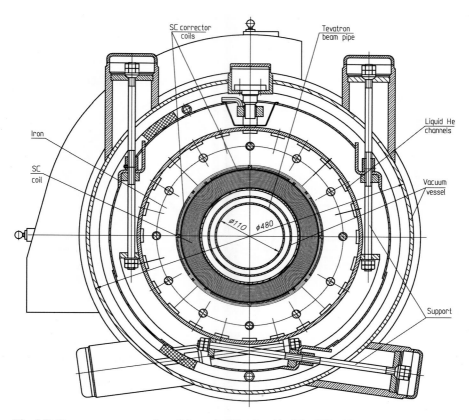

Fig. 2.8 Transverse cross-section of the main SC solenoid of the TEL [39]

is designed to handle 6.5 T, its nominal operating strength is 3.0–3.5 T. During initial operation it successfully reached 6.7 T before quenching.

The gun and collector solenoids use water-cooled copper windings which generate a maximum field of about 0.4 T on the axis with a maximum 340 A of current. The resistance and inductance of the 391 turns of wire is roughly 0.19 Ω and 18 mH. The bore of each magnet has a diameter of about 24.0 cm and a length of 30.0 cm, enough to contain the electron gun and the collector input port. There is a small design difference between the gun and collector solenoids—the collector solenoid has an additional iron plate on its back end, which reduces the field strength outside the solenoid (in the region of the collector itself). Electron beam shape and position correctors are set inside each of the conventional solenoids. The corrector consists of four coils, which can be commutated either as a quadrupole or as two dipoles (vertical and horizontal). Each coil has a layer shape geometry with 0.74° inner and 40.04° outer angles, 11.2-cm inner radius and 0.9-cm thickness. The coil length is 30 cm. The coils are wound from 1-mm diameter copper wire and have 620 turns each. In the dipole configuration, the field is equal to 19 G/A; the

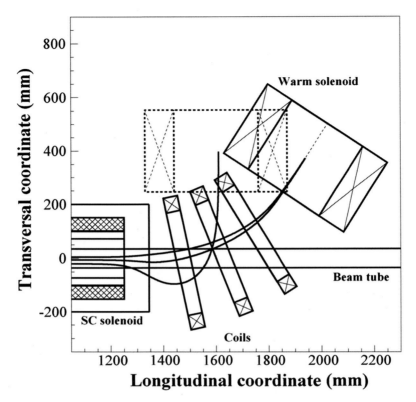

Fig. 2.9 Simulations of the magnetic field lines in the bending sections of the TEL-1 and TEL-2 carried out using the MULTIC code [47]. The placement of the TEL-1 gun solenoid is shown by *dashed lines*; TEL-2 magnets are represented by *solid lines*. Magnetic field lines (electron beam trajectories) in both TELs are shown as well [39]

quadrupole field is equal to 6 G/cm/A. The maximum current in these coils does not exceed 5 A. In routine operation, these corrector coils in the gun and collector solenoids are rarely employed (Fig. 2.9).

The axes of the gun and collector solenoids in very first electron lens TEL-1 were perpendicular to the axis of the main SC solenoid. Operational experience with such a configuration has shown that the electron beam transmission can be assured only within the limited range of the main solenoid field to gun and collector solenoid field ratio $B_{main}/B_{gun} \approx 10$–20 [48]. Beyond this range, the electron beam did not fit the aperture of the electrodes in the bending section of the TEL-1. In addition, there was a significant—several mm—vertical drift of the electron beam due to $\boldsymbol{B} \times \nabla \boldsymbol{B}$ effect in the bending section, which scales as:

$$dy(z) = \int_s \frac{2U_e}{e\beta_e B(z)R(z)} dz, \qquad (2.47)$$

where z is the coordinate along the electron trajectory, U_e is the electron beam kinetic energy, $B(z)$ and $R(z)$ are the magnetic field and the magnetic field line curvature radius. The second electron lens (TEL-2) was designed to significantly increase both $B(z)$ and $R(z)$ in the bending sections, reduce the drift $dy(z)$ 4–5 fold and allow a wider range of the ratios B_{main}/B_{gun} . For that, axes of the gun and collector solenoids (identical to those in TEL-1) were set at 57° with respect to the main solenoid axis and three additional short solenoids were added in each bending section, as shown in Fig. 2.7a and Fig. 2.9. Each of the three new coils generated about 420 G of magnetic field in its center. All the coils were powered in series with the gun or collector solenoids. As the result, the minimum magnetic field in the bending section has been increased from 0.08 T to 0.13–0.18 T for the typical field configuration. The electron beam size in the bending region is reduced as well, as it scales as $a_e(z) = a_{cathode}(B_{main}/B(z))^{1/2}$. Therefore, the ratio of the gun solenoid field to the main solenoid field can now be varied in a much wider range allowing for greater adjustment flexibility of the electron beam size in the interaction region. For example, for $B_{gun} = B_{collector} = 0.3$ T the electron beam can pass the main solenoid with $B_{main} = 0.3–6.5$ T in the TEL-2 while in the TEL-1 the allowed main solenoid field range was limited from 2.7 to 5.5 T.

2.3.1.2   Corrector Magnets

The TEL's ability to adjust the electron trajectory inside the main solenoid to the straight line of the (anti)proton orbit is needed in four degrees of freedom: the upstream position and the angle, both in the horizontal and vertical directions. Six superconducting dipole corrector magnets are used for this steering. Two of these correctors, one oriented horizontally and one vertically, are located at the upstream end of the main solenoid; their goal is to adjust the upstream transverse position of the electron beam to equal that of the (anti)proton orbit. Two other correctors extend nearly the length of the main solenoid. These long correctors have the ability to angle the electron beam along their entire length. Once the upstream correctors are set, the long correctors are adjusted so that the electron beam coincides with the (anti)proton orbit, as drawn in Fig. 2.10. Beam position monitors (described below) installed in the upstream and downstream ends of the long correctors are used to confirm that the two species (electrons and antiprotons or electrons and protons) and are set at identical transverse positions. The electron beam can end at a variety of positions, yet it must be able to pass into the collector. To accomplish this, a third set of correctors are located downstream of the long correctors in order to steer the beam back into a position where it will successfully enter the collector. These correctors, identical to the upstream correctors, often are adjusted simultaneously with either the upstream or the long correctors, but in the opposite direction; in this sense, they "undo" the changes made by the other correctors.

The dipoles are placed on the outer surface of the main SC solenoid coil, as shown in Fig. 2.10. Four pairs of 250-mm long coils form short vertical and

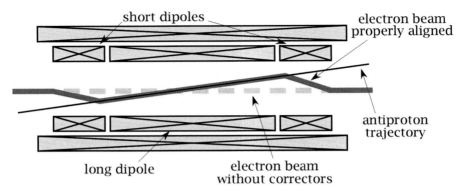

Fig. 2.10 Sketch of the placement and action of the dipole correctors (transverse scale exaggerated). Without activating the correctors, the electron beam would follow the dashed path in the main solenoid. By using the correctors, the electron beam can overlay the (anti)proton path [39]

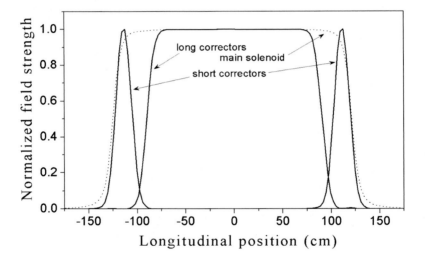

Fig. 2.11 Normalized strength of the dipole correctors ($B_{x,y}/B_{max}$) inside the main solenoid. The solenoid field B_z is also included (*dashed line*). The longitudinal position is referenced to the geometrical center of the main solenoid. The maximum magnetic field is 0.8 T in the short correctors, 0.2 T in the long correctors and 6.5 T in the main SC solenoid [39]

horizontal dipoles at each end of the solenoid. Two pairs of 2-m long coils are placed in the central region of the SC solenoid. The steering dipoles are wound of cable transposed from 8 wires of 0.3-mm diameter. The wire has a critical current of 50 A at 4.2 K and 5 T and Cu/SC ratio of 1.5. The dimensions of the bare cable are 0.45×1.48 mm². The cable is wrapped by polyamide film of 0.03-mm thickness with a 1/3 overlap. The central dipoles have one layer of cable; lateral dipoles consist of two layers and an inter-layer spacer of 0.2-mm thickness. The specific location, and the magnetic length, of each of these correctors is shown in Fig. 2.11.

The dashed line illustrates the main solenoid field on axis as a function of longitudinal position. It is at a maximum nearly from -100 cm to $+100$ cm and rapidly falls to almost zero at -150 cm and $+150$ cm. The solid lines in Fig. 2.11 represent the measured strength of each set of dipole correctors. The upstream short correctors peaks around -115 cm, the long corrector extends from -75 cm to $+75$ cm, and the downstream short corrector peaks at $+115$ cm. Since the strength of each corrector and the main solenoid can be arbitrarily set, their magnitudes are all normalized to 1.0 in Fig. 2.11. In the actual measurements, the solenoid was set to 6.5 T, the short correctors were set to 0.8 T, and the long correctors were at 0.2 T. No significant difference was observed between the horizontal and vertical correctors.

Strongly magnetized electrons spiral around the solenoidal field lines in the TELs. The dipole correctors add a small perturbation to the nominally longitudinal solenoidal field. By superposition, the vector field of the correctors gets added to the vector field of the main solenoid. Since the former is a uniform field pointing transversely and the latter is a uniform field pointing longitudinally, the net result is a field that points at an angle represented by the sum of the two vectors. The electron beam, following the field lines, tracks the resultant field. Beyond the region of the corrector, the field lines and the electron beam again point longitudinally, but from this new position. The total horizontal deflection dx can be derived from:

$$dx = \int_s \frac{B_{horizontal}(z)}{B_{main}(z)}\, dz, \qquad (2.48)$$

where the two field strengths are functions of the longitudinal position z, and the integral covers the pertinent length shown in Fig. 2.11. A similar expression can be written for the vertical corrector, assuming the appropriate corrector field is used. The strength of the short correctors in units of T mm/A, which is $dx \times B_{main}$ per unit current, is about 0.6 T-mm/A. The strength of the long correctors is about 3.6 T-mm/A. Dividing these numbers by the main solenoid strength yields a valid transverse displacement for a known amount of current. For example, short coils energized by 200 A current can move the electron beam by 40 mm in a 3 T main solenoid field, and long coils are able to deflect the beam trajectory in the main solenoid by 30 mrad if energized by 50 A. Separate measurements of electron-beam deflection using BPM readings verified these calibration numbers.

2.3.1.3   Cryogenics and Quench Protection

All the superconducting coils of the main solenoid are immersed in a liquid helium bath, and the total weight of this cold mass is 1350 kg. Due to hysteresis effects and eddy currents in steel, a small amount of heat is generated whenever the current in the superconductor is changed, limiting the maximum current ramp rate to 10 A/s. In practice, the main solenoid is rarely powered up or down and usual ramp rates lie under 4 A/s. The total static heat load onto the helium vessel is 12 and 25 W onto the nitrogen thermal shield of the cryostat. The TEL cryostat is a part of the Tevatron magnet string cooling system which delivers 24 g/s of liquid helium. The magnet temperature margin equals 0.6 K at a helium temperature of 4.6 K.

On rare occasions, the TEL main solenoid has quenched, either on its own or in response to other Tevatron SC magnets quenching first. Since the number of times this has happened is extremely small (about a dozen over 7 years of operation), it is not considered a liability in the performance of the TEL or the Tevatron. Nevertheless, quench protection is an important subsystem of the TEL, as the main solenoid can contain up to 1 MJ of energy when it reaches its maximum rating of 65 kG, and that energy is released over a mere 2 s when the solenoid is quenching. The current in each SC magnet loops through external power supplies, allowing external quench detection circuits and loads to absorb most of that energy. Simulations of quenches suggest that roughly 90 % of the total energy can be dissipated in the external resistive loads, with the remaining 10 % being dissipated in the solenoid itself. In these simulations, the temperature of the hottest point in the coil rises to about 270 K.

The dipole correctors can only contain up to 1.3 kJ of energy, and dissipating this energy within the magnet is not worrisome. However, heat in one region could cause a quench in the main solenoid. Therefore, the correctors are also connected to quench protection circuits and loads.

Each monitor was originally designed to observe the voltage across its magnet and the time-derivative of the current, which were compared to an assigned limiting voltage:

$$\left| L_{magnet} \frac{dI}{dt} - V(t) \right| < V_{limit}, \tag{2.49}$$

where L_{magnet} is the inductance of the magnet. If the difference exceeds $V_{limit} = 1$ V, the magnet is assumed to have begun to quench, and a signal is sent to high current IGBT switches to disconnect the coil from the power supply and allow it to dump the coil current into the resistive load. Mechanical current breakers are installed in series with the solid state switches for redundancy. However, the inductance of a large 0.5 H solenoid is typically not constant at low frequencies (1–10 Hz), due to iron-saturation effects and eddy currents. The overly simplified model expressed in (2.50) led to occasional false quench detections. A more sophisticated model using higher-order effects of both $V(t)$ and $I(t)$ was adopted for operation. The quench protection monitor then tested the relation,

$$\left| L_{magnet} \left(\frac{dI}{dt} + \kappa_1 \frac{d^2 I}{dt^2} \right) - \left(V(t) + \kappa_2 \frac{dV}{dt} \right) \right| < V_{limit}. \tag{2.50}$$

The addition of extra terms in (2.51) offered the ability to better mimic the physical behavior of the magnets over a range of frequencies and dramatically decreased the number of false quench detections.

The power supplies for each of the solenoids and correctors need to be able to sustain each magnet's full current. The main solenoid in normal operation requires a full kilo-Ampere; large MCM500 cables carry this current over some 60 m from the power supply (located in an above ground gallery) to the solenoid itself located in the Tevatron tunnel. The short dipole correctors employ 200-A supplies, while the long correctors use 50-A supplies. Since the correctors might need to be energized in either direction, each supply is fed through a reversing-switch box. This box is able to swap the leads, effectively turning the unipolar supplies into bipolar supplies.

The current ramp rates for each of the superconducting magnets are limited, and all of the settings are done remotely through computer control. The reversing-switch circuits automatically handle ramping the current through zero and switching polarity properly. With this feature, scanning the electron beam transversely was straightforward.

2.3.1.4   Straightness of Field Lines

If there is any significant deviation of the magnetic field lines from a straight line of (anti)proton trajectory, then the electron beam, which follows the field lines, would not interact properly with the (antiproton) bunches. Both initial design considerations and operational experience emphasize the need for the solenoid field straightness within 0.2 mm, i.e., a small fraction of the (anti)proton rms beam size $\sigma = 0.5$–0.7 mm and the electron beam radius $a_e = 1.5$–2 mm in the TEL.

The electron lens magnets were designed and built to be straight and uniform within the specification that was later confirmed in special measurements of the magnetic field lines. This measurement technique is illustrated in Fig. 2.12. A small iron rod was centered in a non-magnetic gimbal equipped with low friction sapphire bearings and mounted on a small cart. The cart was dragged through the solenoid, and the solenoid field magnetized the rod, aligning itself along the field lines (a magnetized ferromagnet feels a torque $M \times B$ attempting to align it along the field lines). A small mirror that was attached perpendicularly to the rod (actually surrounding the rod) reflected a laser beam from one end of the solenoid back down the same direction. The reflected laser beam (returning at twice the angle of the magnetic rod) struck a two-dimensional light-position sensitive detector (PSD). This large PSD was read out by the processing electronics that directly reports the XY coordinates of the incident light spot. In this manner, minute angles $\theta_{x,y} = B_{x,y}/B_{main}$ of the order of few microradians could be observed. The field line coordinates (x,y) are then calculated as:

$$(x, y) = \int_z \theta_{x, y} dz. \tag{2.51}$$

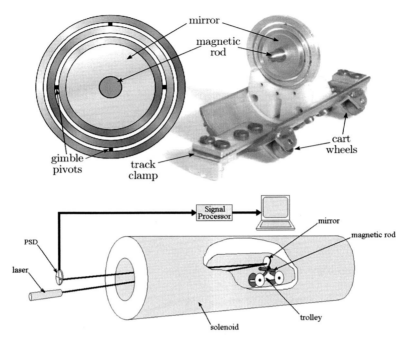

Fig. 2.12 (*Top*) Drawing of the gimble and photograph of the cart. The gimble measured only 2.54 cm across and has very little mechanical resistance. The cart was pulled by a long track, and it rolled inside an aluminum pipe temporarily thrust into the solenoid bore; (*bottom*) simplified cartoon illustrating the technique used to measure the magnetic field line straightness in the main solenoid [39]

The field lines in the center of the solenoid are shown in Fig. 2.13. In the ± 100 cm of the interaction region, the field does not bend more than 200 μm in the horizontal direction and only about 45 μm in the vertical. The rms deviations of the magnetic lines are 15 μm in the vertical plane and 50 μm in the horizontal plane. Therefore the electron beam is able to surround the antiproton bunches through the entire solenoid length. The 200 μm variation is conveniently small and allows the option to experiment with beam-beam compensation at different electron-beam sizes.

The deviation of the straightness of the field lines change as the solenoid's field is ramped up or down is found to not exceed 20 μm. The solenoid field lines as far as ± 1 mm away from each other are measured to be parallel within ± 6 μm [49].

2.3.1.5 Magnetic System of the RHIC Electron Lenses

Though similar, the magnetic system of the RHIC electron lenses has some differences and advances compared to TELs. Its main superconducting solenoid is also a warm bore magnet (154 mm ID) with an operating field up to 6 T. The 2.8 m cryostat includes a number of additional magnets totalling 17 which includes

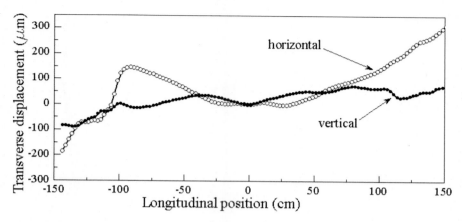

Fig. 2.13 Measured vertical and horizontal field lines in the TEL-1 main solenoid at 4 T [39]

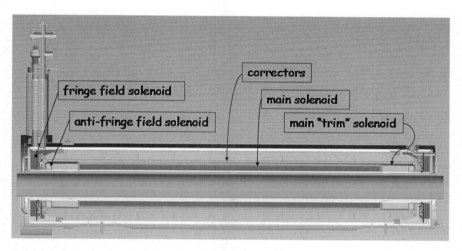

Fig. 2.14 Superconducting main solenoid cryostat of the RHIC electron lens contains main solenoid, fringe field and anti-fringe solenoids, and a number of the field straightness and angle correctors [41]

besides the main coil, two fringe field coils, two anti-fringe coils, 5 vertical and 5 horizontal dipole correctors to guarantee the main field straightness, and two angle dipole correctors for beam positioning, all operating at 4.2 K—see Fig. 2.14. The fringe field solenoid coils at both ends are included to assure greater than 0.3 T guiding and focusing solenoid field for the electrons between the superconducting magnet and the warm transport solenoids GSB and CSB (see Fig. 2.7). To achieve the desired field uniformity over a range of field strengths B_{main}, anti-fringe field coils are placed next to the FF coils. All these coils at both ends of the magnet can

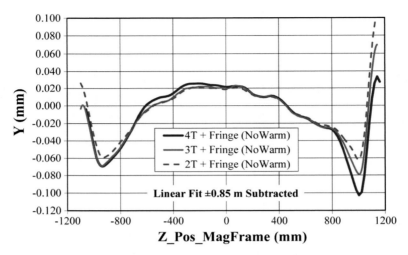

Fig. 2.15 Vertical solenoid field straightness measurement in the 6 T SC main solenoid of the RHIC electron lens (*Yellow lens magnet*). For comparison, the rms proton beam size at 250 GeV is as about 300 μm [40]

be powered independently to avoid forming a "magnetic bottle" with a low main solenoid field, which would trap backscattered electrons.

Five short (0.5 m) dipole correctors in both the horizontal and the vertical planes are installed to correct the solenoid field straightness to ±50 μm. The straightness of the solenoid field lines has tight tolerances (±50 μm over a range of ±800 mm) to ensure good overlap of the hadron and electron beam. The "magnetic needle and mirror" measurement system similar to the one used for TELs showed that even without the SC correctors, the field lines are straight within the specs, so no further correction was found necessary—see Fig. 2.15.

Two 2.5 m long dipole correctors (one for each transverse plane) control the angle of the electron beam inside the main magnet to be changed by ±1 mrad at the maximum field of 6 T, and therefore, to align the angles of electron and proton beams. Positioning electrons onto the protons is provided by horizontal and vertical steering magnets placed within 0.45 T warm solenoids GS2 and CS2 which can move the electron beam in the main solenoid in either plane by ±5 mm.

2.3.2   Electron Beam System

2.3.2.1   Electron Guns

The magnetic system of the TEL allows adiabatic magnetic compression of the electron beam cross section area by a factor of $(a_c/a_e)^2 = B_{main}/B_{gun} \approx 10$. In order to have the electron beam radius a_e be several times the rms (anti)proton beam size

σ in the TEL, the cathode radius should be $a_c = 5\text{–}10$ mm. The requirement of bunch-by-bunch current variation for the long-range beam-beam compensation in the Tevatron calls for the electron beam of the electron lens to be modulated with a high duty factor and a characteristic on-off time of about 0.5–1 μs. The high current density, fast modulation and the requirement of a smooth current density profile led to a choice of an electron gun with a convex cathode which allowed higher perveance and current modulation by the anode voltage (i.e., no grid). During experimental beam studies and operation of the TELs several electron current profiles were found to be effective: (a) a rectangular distribution results in a uniform tune shift dQ for all high energy particles passing through the electron beam, but has the disadvantage of strong nonlinear space charge forces beyond boundaries of the electron beam; (b) a bell-shape (close to a Gaussian) distribution has weaker nonlinearities but a smaller beam size as well, that makes it somewhat cumbersome to align it on the beam of (anti)protons; (c) a "smooth edge and flat top" (SEFT) distribution combines the advantages of both previously mentioned distributions; (d) hollow electron beam distribution for transverse collimation.

Correspondingly, four electron guns have been developed for the TELs. One of the most important characteristics of an electron gun operating at the space charge limit is its perveance P:

$$P = J/U_a^{3/2}, \tag{2.52}$$

where J is the beam current and U_a is the anode potential with respect to the cathode. For the guns with flat or concave cathodes, current density inhomogeneity becomes large when the perveance exceeds the value of 1–2 μA/V$^{3/2}$. In the case where the gun has to be immersed into a strong longitudinal magnetic field, the perveance can be increased by usage of a convex cathode [20].

The electron guns were simulated and optimized using UltraSAM code [50] in order to have the desired current density distribution and high perveance. The geometries of three guns are shown in Fig. 2.16 together with the electric field distribution along the beam axis (the guns have axial symmetry) and the electron trajectories. The guns employ spherical cathodes with a ±45° opening angle. A Pierce-type electrode (control electrode) is installed around the cathode of the "flat" gun (Fig. 2.16a) for manipulation of the beam current density distribution. Control electrodes of different geometry are installed for the same purpose between cathode and anode in the "Gaussian" and around the cathode in the SEFT guns (Fig. 2.16b, c). The control electrodes in the latter two guns are usually kept at the same potential as the cathode.

Mechanically, all three guns look similar to what is presented in Fig. 2.17a. They are assembled on a 171.5 mm (6 ¾ in.) diameter stainless steel vacuum flange and use ceramic rings as insulators between electrodes. The guns employ spherical convex dispenser cathodes purchased from HeatWave Labs (Watsonville, CA). The 10 or 15 mm diameter barium-impregnated tungsten cathodes operate at temperatures of 950–1200 °C. They are equipped with a Mo-Re support sleeve and

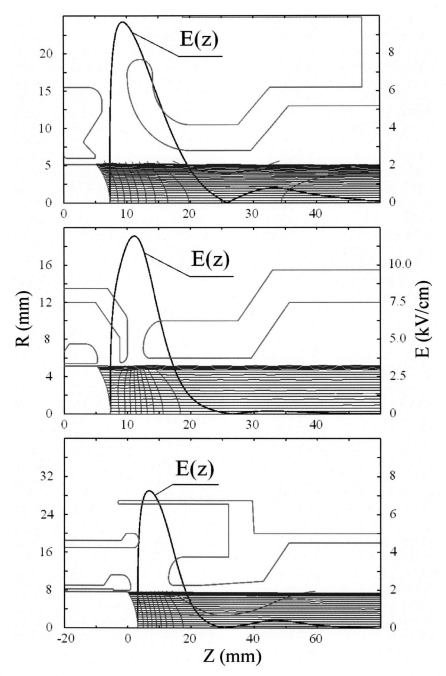

Fig. 2.16 Gun geometry and "UltraSAM" code electric field simulation results for TELs: (**a**) "flat-top" gun; (**b**) Gaussian gun; (**c**) SEFT gun [39]

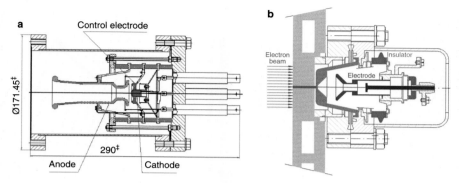

Fig. 2.17 (**a**) Mechanical design of the TEL "flat-top" gun; (**b**) pin-hole collector assembly for beam profile measurements on the test bench [39]

molybdenum mounting flange and have an internal heater filament (bifilar option, one heater lead internally grounded). The near-cathode electrodes are made of molybdenum, while the control electrodes and anodes are made of oxygen free copper.

Gun characteristics were measured on the test bench used at Fermilab for prototyping the TEL elements [51]. The test bench consists of the gun immersed into a longitudinal magnetic field B_{gun} of 0.1–0.2 T generated by a gun solenoid, a drift tube with diagnostics placed inside a 0.4 T, 2 m long main solenoid, and a collector, also inside a separate solenoid. The collector is equipped with a beam analyzer, illustrated in Fig. 2.17b. A small 0.2 mm diameter hole in the collector base lets a narrow part of the electron beam to pass through a retarding electrode and be absorbed by an analyzer collector. To measure the transverse current density distribution, the beam is moved across the hole by the steering coils installed inside the main solenoid, and the analyzer collector current is recorded as a function of the transverse beam position.

Except for their high perveance, the guns are not much different from a planar cathode gun. The beam currents follow the Child's law with a good precision (Fig. 2.18) yielding perveances of 5.3, 4.3, 1.8 $\mu A/V^{3/2}$ for the "flat-top", SEFT and "Gaussian" guns, respectively. To prevent thermal problems at the collector, the total current and profile measurements were done in the DC regime at currents below 0.5–1 A. The gun characteristics at higher currents were investigated in a pulsed regime with the pulse width of 0.2–4 μs. No significant deviation from the results of the DC measurements was found.

An example of the 2D profile of the electron current distribution from the SEFT gun measured by the "pin-hole" collector is shown in Fig. 2.19a. Current density variations are less than 10 % over 90 % of the beam diameter. Measured and calculated one dimensional profiles of electron beams from all three guns are presented in Fig. 2.19b. There is good agreement between predicted and observed current densities over most of the beam area except at the very edge of the beam.

Fig. 2.18 Volt-Ampere characteristics of the three electron guns for TELs, solid lines are fits according to Child's law $P = I/U_a^{3/2}$ [39]

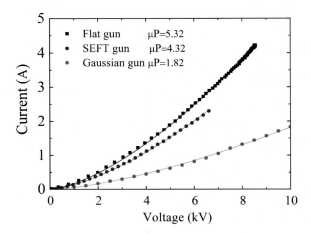

Fig. 2.19 (a) 2D electron current density distribution of the SEFT gun beam; (b) 1D current density distributions for three guns, solid lines represent UltraSAM simulation results. In both cases, the control electrode voltage was set equal to the cathode voltage [39]

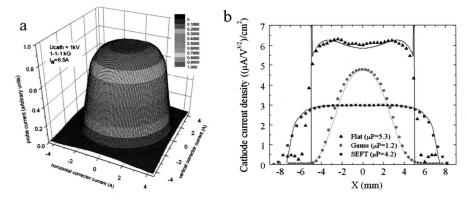

Electron emission from the edges of the cathode is strongly dependent on the accuracy of the mechanical alignment of the near cathode or the control electrodes w.r.t. the cathode. Figure 2.20 shows a 1D current profile in the case when the control electrode of the flat gun was (unintentionally) set a bit farther from the anode than the cathode. The edge peaks in the current density profile indicate, and computer simulations confirm, that the reason is some 0.4 mm protrusion of the emitting surface from the near-cathode electrode with respect to its optimum position. The shift occurred because of either uncertainty in the thermal expansion of the cathode or mechanical error. Probably, a slight current distribution asymmetry, seen in Fig. 2.20, is because of an asymmetric misalignment. The application of negative voltages to the near-cathode electrode (with respect to the cathode potential) resulted in suppression of the electron emission at the edge and can lead to a

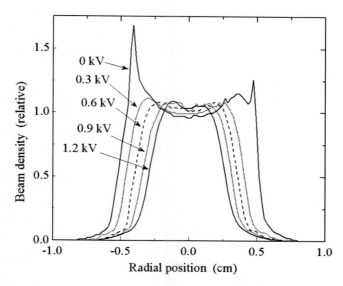

Fig. 2.20 The electron current density profile generated by the "flat" gun at different voltages U_{pr} on the near-cathode electrode. The anode-cathode voltage $U_{ac} = 3$ kV and the magnetic field in all solenoids is 2 kG [39]

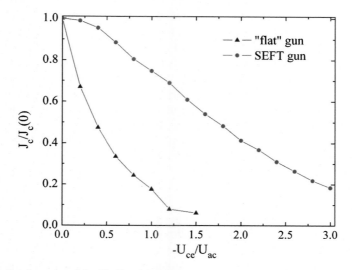

Fig. 2.21 Total current of the "flat" and SEFT guns vs. profile controlling voltage [39]

narrower (almost bell-shape) current profile, as illustrated by Fig. 2.20. Total beam current reduction factors for the SEFT and the "flat-top" gun are shown on Fig. 2.21 as functions of the (negative) profile control voltage U_{pr} normalized to the (positive) anode-cathode voltage difference U_{ac}.

Table 2.3 Main parameters of the TEL electron guns

Parameter	Gun #1	Gun #2	Gun #3	Gun #4	Units
Cathode diameter	10	10	15	15	mm
Current profile	Rectangular	Gaussian	SEFT	Hollow	
Gun perveance, max	5.9	1.7	4.2	3.1	$\mu A/V^{3/2}$
Max. current density	6.3	4.8	3.0	2.8	$\mu A/V^{3/2}/cm^2$
Control voltage to shut off	2.5	3	3.5	n/a	$U_{control}/U_{anode}$
Filament power	35–45	35–45	60–70	60–70	W
B-field on cathode	1–4	1–4	1–4	1–4	T

The total filament power required to keep the cathode at an operational temperature of 1000–1100 °C is about 35–45 W for the 10 mm diameter cathode (used in the "flat" and "Gaussian" guns) and about 60–70 W for the 15 mm diameter cathode used in the SEFT gun (see Table 2.3). For initial activation of the cathode, the power is increased by 30–50 % for a short period of time until the cathode starts to generate enough current to follow the Child's law (2.53) at the design cathode–anode voltages. Special care is taken in order to have sufficiently good vacuum (better than 10^{-8} Torr) in the gun area in order not to poison the cathode which may result in reduced cathode lifetime. With all these precautions, the cathodes of the guns installed in the TELs operated for several years without significant deterioration. If the gun is exposed to air at high cathode temperatures, the cathode is ruined (the tungsten gets oxidized, creating a layer with a high work function) and either a complicated cathode surface processing or (easier) a cathode replacement is needed.

The RHIC electron gun design—see Fig. 2.21—follows a similar concept to provide a beam with a transverse profile that is close to Gaussian [41]. Its cathode radius of 4.1 mm gives a Gaussian profile with 2.8 rms beam sizes. The perveance of the gun is 1.0 $\mu A/V^{3/2}$, giving a total current of 1 A at 10 kV; the current density of the electron beam on its radial periphery can be changed with the control electrode voltage, while the general shape of the beam profile remains Gaussian. The cathodes (LB_6 and IrCe) were produced at BINP in Novosibirsk. With a nominal current density of 12 A/cm², IrCe was chosen as the cathode material for its long life span (greater than 10,000 h). The gun and collector vacuum is UHV compatible, all of the components are bakeable to 250 °C to attain a design pressure of 10^{-10} Torr (Fig. 2.22).

A 15-mm-diameter hollow electron gun was designed, built and installed in one of the Tevatron electron lenses for the purpose of the transverse beam collimation [12]. It was required to have no electrons inside the hole and sharp edges, so the gun was based on a tungsten dispenser cathode with a 9-mm-diameter hole bored through the axis of its convex surface. The peak current delivered by this gun was 1.1 A at 5 kV. The current density profile was measured on a test stand by recording the current through a pinhole in the collector while changing the position of the beam in small steps. A sample measurement is shown in Fig. 2.23. The electron current of the gun was pulse and could be synchronized with practically any bunch or group of bunches in the Tevatron [52].

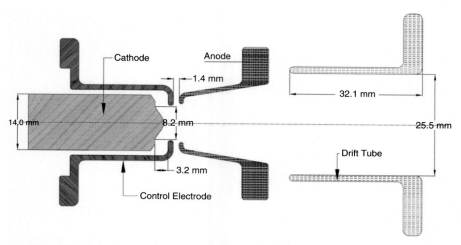

Fig. 2.22 Schematic of the electron gun for RHIC electron lens [40]

2.3.2.2 Electron-Beam Collector

A sketch of the TEL electron beam collector is shown in Fig. 2.24, with a view of the expanding electron beam. Spreading out the beam has two advantages, one being the distribution of the heat load over a larger area and another is containing secondary electrons in the collector. Locating the heating in one spot could melt the copper if the TEL is operated at full beam power (3–5 A at 10 kV). Instead, the magnetic field lines beyond the collector solenoid spread out, and the electrons, following the field lines, are absorbed by a much larger area of the copper collector. Additionally, chilled water is piped into the collector, where it passes through ducts within the copper and extracts up to 50 kW of heat. The intake pipe is shown in Fig. 2.24 to illustrate the setup. Another serious concern is the production of secondary electrons whenever an energetic primary electron impinges on the surface. The secondary electrons penetrating back into the main solenoid can adversely interact with the primary electron beam, generating a two-stream instability even if magnetized [27].

The design of the collector targets this issue by employing the "magnetic bottle" principle—only electrons with a small enough transverse velocity can travel from a region with low-magnetic field $B(0)$ to higher magnetic field $B(z)$:

$$v_\perp^2 < \frac{v_\parallel^2}{\left(\frac{B(s)}{B(0)} - 1\right)}. \tag{2.53}$$

An electron must originate with enough parallel momentum to overcome the magnetic compression; if the electron does not have enough momentum, it will

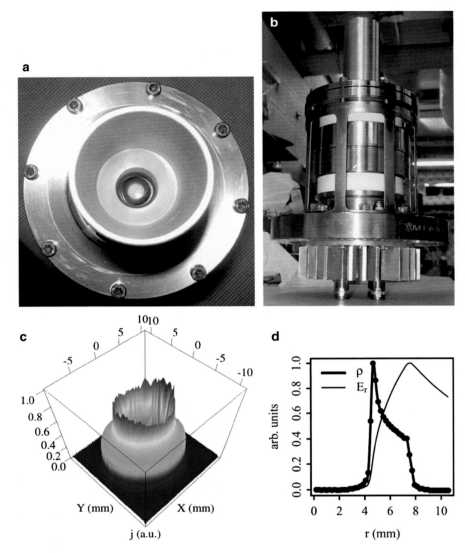

Fig. 2.23 Hollow electron gun: (**a**) top view; (**b**) side view; (**c**) measured current density profile; (**d**) measured charge density $\rho(r)$ and calculated radial electric field $E_r(r)$ [12]

run out of longitudinal momentum and be returned back to the collector surface. The TEL's collector surface has a residual field on the order of $B(0) \approx 0.01$ T, while the collector solenoid runs at 0.38 T, implying that the longitudinal momentum needs to be over six times larger than the perpendicular momentum. If the electrons are emitted from the collector surface uniformly over all solid angles, only 1.2 % of those electrons meet that constraint. More importantly, the number of electrons that

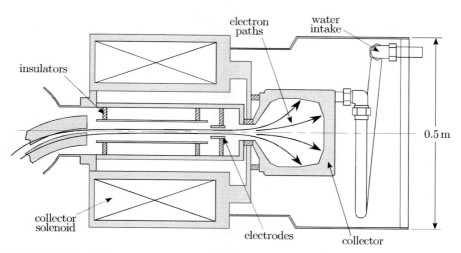

Fig. 2.24 Scaled sketch of the collector cross-section. The collector itself is a water-cooled copper cavity that resides outside the solenoid. This allows the electron beam to spread out, distributing the heat load and decreasing the production of secondary electrons [39]

can pass all the way into the 3.5 T main solenoid is 0.14 %. Experimentally, by comparison of the cathode and the cathode currents (explained in the next section), we determined that under normal operation conditions the collector is able to retain at least 99.7 % of the incident electron beam.

The collector is electrically isolated from the rest of the system. The electron-beam current absorbed by the collector is brought back, via a floating power supply, to the cathode (this is detailed in the next section). The power supply voltage is positive and adjustable. Therefore, the electrons are born at the negative cathode potential and accelerated by the anode and the beam pipe, but as they approach the collector, they are slowed down to a potential somewhat more positive than the cathode one. One of the advantages of this recuperation scheme is that the heat load generated by the incident beam is directly related to its kinetic energy with respect to the collector: e.g., a 10 keV electron on a −6 kV collector imparts only 4 keV of energy. When the TEL is operated at maximum beam power, the total power deposited in the collector could be considerable, and lowering the voltage difference is very useful.

A test of the collector consists of setting its voltage to nearly that of the cathode and measuring the current that it receives. The data from this experiment is shown in Fig. 2.25. At zero voltage difference, only one-quarter of the beam can still manage to reach the collector surface. However, as the voltage is raised, the amount of current reaching the collector increases. As the difference approaches 1 kV, all of the current is received.

Using the data point where the maximum current is truly witnessed, the acceptance of the collector appears to be between 9.2 and 13.6 μP. The collector never limited the TEL's performance, since its voltage could always be increased if

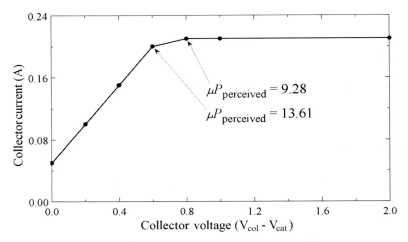

Fig. 2.25 Measurement of the collector acceptance. As the collector voltage is adjusted with respect to the cathode voltage, the current admitted to the collector changes [39]

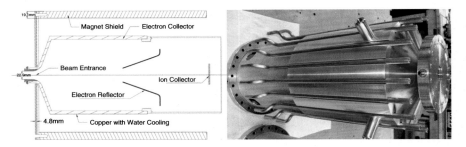

Fig. 2.26 RHIC electron lens electron beam collector: schematic (*left*) and photo (*right*) [40]

necessary. Most commonly, the collector voltage is set at about 2–5 kV above the cathode voltage. Two electrodes, which have primarily been used for monitoring the beam's passage into the collector are also shown in Fig. 2.24. For example, the "scraper" electrode, closest to the collector, has its own current monitor, and if it reads something other than zero, the TEL beam is, at least partially, running into it. Adjusting the downstream magnetic correctors such that this signal is eliminated assures optimal performance of the TEL.

Similarly, the design of the collector for the RHIC electron lens [53] was dictated primarily by the UHV requirements of the collider. It separated the heavily electron-bombarded area from the rest of the electron lens by using a small diaphragm. A magnetic shield leads to fast diverting electrons inside the collector. The collector spreads the electrons on the inside of a cylindrical surface that is water-cooled on the outside (see Fig. 2.26). Simulations give a power density of

10 W/cm^2 for a 10 A electron beam, decelerated to 4 keV, while such a collector can absorb up to four times this power density. The reflector electrode has a potential lower than the cathode and pushes electrons outwards to the water-cooled cylindrical surface. Under a load twice as high as expected from a 2 A electron beam, the maximum temperature on the inner surface of the shell is about 100 °C, i.e., quite acceptable for the material (copper) and for UHV conditions in RHIC.

2.3.2.3   Electric Circuit

A recirculating electron beam could be generated by the simple circuit illustrated in Fig. 2.27. In this case, the cathode power supply does not need to produce any current. The beam current flows through the collector power supply, but the voltage that this supply must support can be significantly less. The anode modulator generates short positive voltage pulses but carries no DC current and simply "kicks" the current around the loop (see discussion in the next chapter).

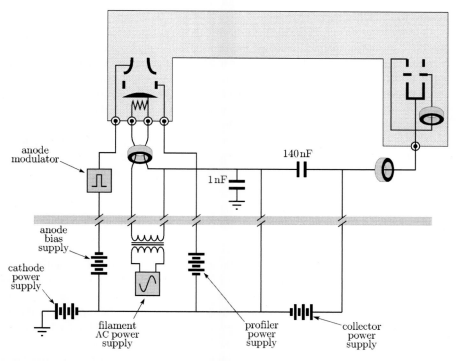

Fig. 2.27 Schematic of the TEL electrical circuitry. The shaded barrier represents about 60 m between the Tevatron tunnel and the gallery where supplies and electronics are stationed. The filament is at cathode potential, so the AC power supply providing the cathode heating needs to be isolated from ground via a transformer, and its signal needs to be subtracted from the cathode current meter [39]

The Tevatron tunnel, filled with beam induced radiation during operation, is of very limited access and a poor place for solid-state electronic equipment. Therefore, the power supplies must be located above the tunnel, connected by cables over 60 m in length. For the majority of these cables, high-voltage, shielded coax (RG-213) of 50 Ω impedance is used. The filament is powered by 60 Hz AC which is electrically isolated through a high-voltage transformer. This allows the primary side to be referenced to ground and only the secondary is at the cathode voltage. The electron current from the cathode is measured by a high-bandwidth, commercial current transformer that encircles the wire attached to the cathode. The transformer's signal is preserved over the long Heliax cable by only grounding it upstairs where an oscilloscope measures it. This prevents ground noise from corrupting the signal. Keeping the impedance at 50 Ω also reduces electrical coupling from other sources over the long propagation distance and reflection issues. Rise-times of 1–2 ns and currents of a few mA are visible. However, the low-frequency filament current passes through the current transformer, so the return cable was obliged to pass through it also; the two currents are always in opposition and therefore cancel each other. In order to house and connect the capacitors, current meters, and make additional interconnections, a high-voltage enclosure was constructed and installed next to the electron gun.

In order to confirm that the electron beam arrives at the collector without losses, another current transformer monitors the current returning from the collector to the recirculating capacitor. A third transformer watches the scraper electrode's current since this provides the narrowest aperture and the easiest way to adjust field strengths in order to steer the beam into the collector. The scraper feeds into the collector cable, so that the collector current should be identical (though delayed) to the cathode current. Above several hundreds of mA, the peak collector current is somewhat less than that of the cathode current due to electron pulse lengthening induced by the electron's own space-charge in the beam pipe.

The electric circuitry of the RHIC electron lenses is very similar to the Tevatron one [40].

2.3.2.4 Electron-Beam Modulators

Beam-beam effects are unique for each of the Tevatron proton or antiproton bunches, correspondingly, compensation of them requires different electron currents be prepared for each bunch. There are 1113 RF buckets along the Tevatron orbit (RF frequency 53 MHz, revolution frequency 47.7 kHz), but only 36 of them are populated with proton bunches and 36 with antiproton bunches. For each kind of particles, the populated buckets are arranged in three trains with twelve bunches in each of them. The total length of each bunch train is approximately 4.5 μs. The distance between the batches is 2.6 μs. The bunch spacing within a train is 396 ns. The interaction length of the TEL is about 2 m and it takes 33 ns for 10 kV electrons ($\beta_e = 0.2$) to traverse its length. The (anti)proton bunch enters the drift space of the lens when it is already filled with electrons. For (anti)protons ($\beta = 1$) it takes 6 ns to

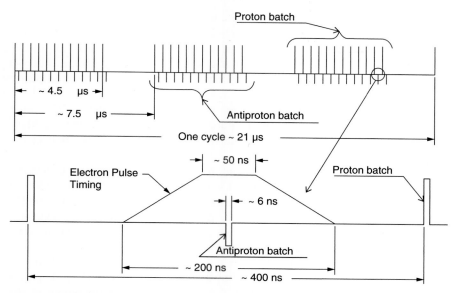

Fig. 2.28 The Tevatron beam structure and required TEL electron pulse structure [39]

pass through the TEL. After the last particle of the (anti)proton bunch leaves the drift space, it is possible to shut off the electron gun. Taking into the account also the (anti)proton bunch length of about 10 ns, the minimum flat top length requirement to the TEL gun's extraction voltage pulse is about 50 ns. The Tevatron bunches share the same beam pipe and, ideally, in order for the TEL to act on only one type of particle leaving another unaffected, the total electron pulse duration should be limited to less than 400 ns. The proton and antiproton bunch structure in the Tevatron and the required timing of the electron beam in the TEL are shown in Fig. 2.28. Besides the pulses of electron beam to compensate for beam-beam effects, additional electron pulses are required in the abort gaps between the bunch trains (not shown in the picture) for the TEL to remove uncaptured DC beam particles. Therefore, the system ideally should produce 39 pulses during each period of particle revolution in the Tevatron (every 21 μs). The required electron current modulation can easily be done by a gridded gun, but such a method is not acceptable for the TELs as it destroys the carefully prepared electron current density distribution generated on the cathode. Therefore, the modulation should be done by full anode voltage of 5–10 kV and that results in extremely challenging requirements for the TEL gun HV modulator. The requirements have been significantly eased by following three considerations: (a) the proton and antiproton beams are spatially separated by some 8–10 mm in the area where TEL is installed and it was found that electron beam centered on one of beams produces very little harm on the other one—therefore, the total required electron pulse length can be as long as 800 ns (to fit between three neighbor bunches of the same kind); (b) only a few (3–6) bunches of protons suffer greatly from the beam-beam effects and need immediate compensation by the lenses and consequently, only few electron pulses per turn are needed; (c) finally, it was found that functions of the two TELs can be separated—one of them could be used for the DC beam cleaning in the abort gaps while another can do beam-beam compensation—thus the need to generate simultaneously many different purpose electron pulses dropped.

Several different types of high-voltage anode modulators for the TELs were developed and tested [54–57]. The RF-tube based amplifier [54] and the Marx generator [56] were found most suitable for the electron lens operation in the Tevatron.

The first modulator type uses the output from the anode of a grid driven tetrode. The tube anode is connected to a +10 kV DC anode supply through a 1500 Ω resistor. The modulating voltage on the anode of the tetrode is then AC-coupled through two 1000 pF ceramic capacitors to the electron gun anode. This modulator has the advantage that it is not susceptible to radiation damage and can be installed directly adjacent to the Tevatron beamline. A CPI/EIMAC 4cw25000B water-cooled tetrode, with a maximum plate dissipation of 25 kW, is used in this modulator. Its anode voltage is supplied by a Hipotronics 10 kV, 16 A, dc power supply. An additional LC filter (1.5 H, 20 μF) was added to the output of the Hipotronics supply to reduce ripple to less than 1 part in 10,000. The anode supply is connected to the tetrode through a 1500 Ω, 250 kW, water cooled resistor (Altronic Research). The grid of the tetrode is driven by an IGBT pulser. For compensating a single bunch of protons or antiprotons, the tube is typically operated with a screen voltage of 500 V and a DC grid voltage of 0 V. The tetrode's grid is then pulsed with a negative voltage pulse from the IGBT pulser, reducing the current flow through the tetrode. The positive pulse appearing on the anode is then coupled, using two 1000 pF ceramic capacitors in parallel, through a short (0.6 m) section of 50 Ω, RG213 cable to the anode of the electron gun. Since the gun anode must be charged through the 1500 Ω resistor, the rise-time is limited by the sum of the tetrode's anode-screen capacitance (35 pF), the capacitance of the cable connecting the modulator to the gun (60 pF), and the gun anode to ground capacitance (60 pF). Typical rise- and fall-times are ≈300 ns and total output pulse duration from such a modulator is 800–1200 ns [54]. A pulse to pulse amplitude stability of 0.02 % was achieved by applying a feed-forward compensation signal to the grid of the tetrode to reduce ripple on the modulator output at power line frequencies. The RF-tube based modulator is in routine use in the TEL-1 since 2001.

The solid-state Marx generator drives the anode of the electron gun to produce the electron beam pulses in the second TEL. It drives the 60 pF terminal with 600 ns pulses of up to 6 kV with a repetition rate of 47.7 kHz and with rise and fall times of about 150 ns. The generator consists of capacitor banks charged through inductors and erected using triggered switches, typically solid-state IGBTs. Stangenes Industries constructed ours, and each of its 12 stages can be charged to 1.2 kV and erected in series to produce a 14 kV pulse [57]. The discharge IGBT switches erect the pulse with a rise time of 150 ns, and the charging IGBT switches terminate the pulse

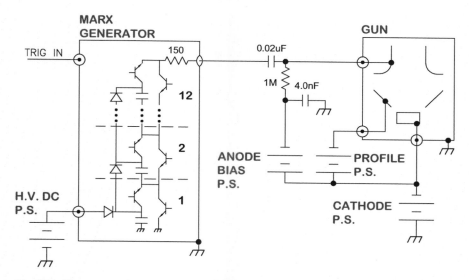

Fig. 2.29 Electron gun driving scheme. Only the cathode, the control electrode (profiler) and the anode of the electron gun are shown [39]

with a fall time of 150 ns, and then provide a path for charging the capacitors in parallel.

A major design issue in the system has been the excessive switching losses in the IGBT's while running at approximately 50 kHz rep rate. The device is air-cooled, and the system has run into thermal problems that have limited it to operating at less than 6 kV at the nominal rep rate. The Marx generator was designed to drive approximately 60 pF, so it must be mounted in close proximity to the gun anode terminal to avoid the extra capacitance of a long connecting cable (see Fig. 2.29).

TEL2 is located in the Tevatron tunnel, a few yards downstream from the collider beam dump. This is arguably the worst location in the tunnel for solid-state equipment that is not radiation-hardened. The Marx generator stopped functioning in less than a week of Tevatron operation, and we found it to be Class I radioactive when we removed it from the tunnel. However, after cooling down for a few days, it started operating again. It was reinstalled behind two feet of steel shielding, and it functioned for several weeks before failing. Four more feet of shielding were added, and the unit has been operating continuously for almost a year.

Figure 2.30 shows the output voltage of the Marx generator when gated with a 520 ns pulse. At 4.2 kV of Marx output voltage and anode bias voltage of 200 V the peak electron current is 1 A which is consistent with the SEFT gun perveance.

A second-generation Marx generator with water-cooled IGBT's with 12 different voltage levels (during one cycle, train) was developed later in the Tevatron Collider Run II [58]. It was able to work at higher voltages at the 50 kHz rep rate and a solid-state modulator based on summed pulse transformer scheme to generate similar

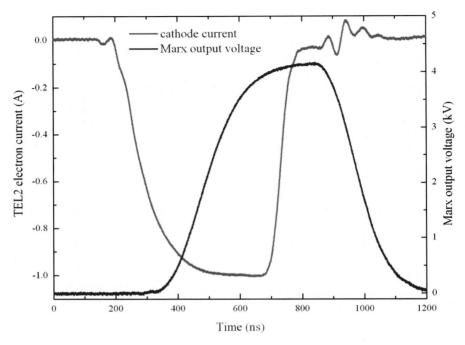

Fig. 2.30 TEL-2 Marx generator output voltage and electron beam current [39]

waveforms, but at a rate of 150 kHz, i.e., suitable for long-range beam-beam compensation for all (anti)proton bunches.

2.3.3 Beam Diagnostics and Other Sub-systems

The electrons streaming out of the gun in the electron lens travel through a beam pipe until they are finally absorbed in the collector. The beam pipe through which the electron beam travels preserves a constant inner diameter along nearly the entire length of the electron-beam path. This starts just downstream of the anode, continues around the first bend, through the main solenoid, around the second bend, and into the collector solenoid. The inner pipe diameter in the main solenoid needed to be as large as the physical aperture of the Tevatron, as the TEL was not intended to ever inhibit the performance of the Tevatron. The dotted line in Fig. 2.31 outlines the typical aperture found around the Tevatron ring. The circumscribed circle of radius 35 mm corresponds to the inner surface of the pipe through the entire length of the TELs. The TELs' vacuum components were certified according to standard Tevatron vacuum requirements including cleaning and vacuum baking. An in-vacuum heater was installed to achieve the necessary in-situ baking temperature after the assembly and installation. Three 75 l/s ion pumps and a TSP were installed

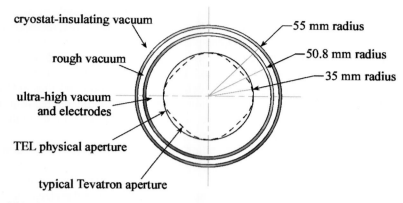

Fig. 2.31 Cross-section of the TEL beam-pipe [39]

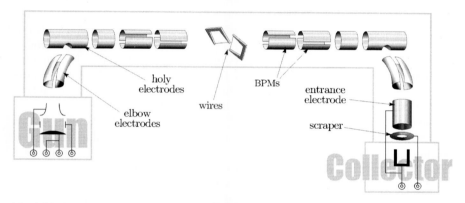

Fig. 2.32 Sketch of the electrodes in the TEL (not to scale) [39]

to maintain the ultra-high vacuum in the system. The TEL vacuum under working conditions ranges from 8×10^{-10} to 3×10^{-9} Torr that is comparable with the gas pressure in the nearby sections of the Tevatron. The vacuum valves at both ends ensure that the system can be separated from the Tevatron vacuum to perform maintenance if needed.

All the elements of the TEL interaction region have the same inner diameter of 70 mm as the adjacent Tevatron beam-pipes in order to minimize machine impedance. Figure 2.32 sketches the numerous electrodes involved in detecting and measuring the electron beam and the antiproton and proton bunches. Each of these electrodes is electrically isolated from the grounded beam pipe and is wired, through vacuum feed-throughs, to coaxial cables leading out of the Tevatron tunnel and to the support electronics. The wires shown in the center are two mechanically actuated forks. One is oriented vertically, the other horizontally, and each has a 15 mm long 0.1 mm diameter tungsten wire strung across the gap.

Remotely operated motors are able to swing each fork into the middle of the beam pipe, where the electron beam is flowing. By adjusting the correctors, the beam can be swept across the fork, and the intercepted charge flows through the fork and again into cables that bring the signal out to be measured. The amount of current as a function of beam position yields data that can be converted into a profile of the current density using Abel inversion (see details and the resulting profiles in [59]). The forks are always positioned outside the beam pipe before stores begin in the Tevatron. The second TEL does not have these forks as we were more confident with the generation of the electron current profiles.

The elbow electrodes are curved imitating the path of the electron beam around each of the bends. Horizontally opposed, a high-voltage difference can be applied in order to generate a strong horizontal electric field to create a small vertical drift of the electron beam if needed, but during normal TEL operation the electrodes are grounded.

The holey electrodes are simply cylindrical electrodes that have a hole cut into one side. The electron beam passes through this hole as it enters and leaves the region of the (anti)proton orbit. These and the elbow electrodes were installed to assist with initial TEL commissioning. If the electron beam failed to pass through the solenoids and into the collector, observing which electrodes were absorbing the electron current would indicate how to correct the guiding fields. Fortunately, the electron beam had little difficulty propagating completely into the collector, and the utility of these electrodes diminished quickly.

Next to the holey electrodes in Fig. 2.32 are cylindrical electrodes intended for clearing out ions. Ions are created by electrons bombarding residual gas molecules floating in the beam-pipe vacuum. The once-neutral molecule can easily lose electrons, turning it positively charged and attracted to the electron beam's space charge. In actuality, the influence of ions has been small enough not to induce instabilities or other problems, and these electrodes are typically grounded.

The TEL is equipped with four beam-position monitors (BPMs): one vertical and horizontal at the beginning and at the end of the main solenoid. The BPMs are pairs of plates that pick-up signals when any charged particle passes through them. Figure 2.33 shows an example of the voltage seen on one of the BPM plates during passage of the electron pulse and few proton and antiproton bunches.

The average position of any beam charge can be calculated as:

$$x = k \frac{V_A - V_B}{V_A + V_B}, \tag{2.54}$$

where x is the transverse distance of the beam from the center of the beam pipe, $V_{A,B}$ are the measured voltages on the two plates, and $k = 33.6$ mm is a constant empirically determined from calibration measurements. To reduce the noise, each V_i is calculated as the total integral of the bunch's charge profile, which in turn is the integral of the doublet current signal as charge is first pulled onto the electrode, then returned, as the bunch passes by [60].

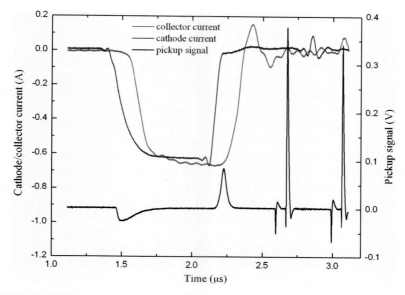

Fig. 2.33 BPM pickup signal (*black*) featuring the electron pulse, two proton (positive peaks) and two antiproton bunches (negative peaks). For clarity the electron pulse is shown timed to the abort gap where no bunches are present. The scope bandwidth was intentionally set to 20 MHz to reject high frequency noise and limit proton bunch signal amplitude. Cathode (*blue*) and collector (*red*) currents are measured by current transformers. The artifacts on the right hand side are caused by the instrumentation [39]

Each 20-cm long BPM measures the beam's position in only one dimension and at only one longitudinal position. Therefore a pair of BPMs is needed on the upstream end of the main solenoid in order to record both the horizontal and vertical positions, and another pair of BPMs are needed on the downstream end. In the TELs, a fast Tektronix TDS520 oscilloscope and a LabVIEW application program operating on a dedicated PC computer process the waveform signals and compute positions constantly during stores. The design of this system includes the ability to measure the position of antiproton bunches and proton bunches with the same BPMs plates as the electron beam. In this manner, it becomes possible to confirm the electron beam and the (anti)proton bunches are collinear within the main solenoid.

The four position detectors are sequentially connected to the oscilloscope's inputs through the Keithley RF multiplexer. The computer communicates with the oscilloscope and the multiplexer through a GPIB interface and links with the Tevatron accelerator control net through Ethernet. A beam synchronous pulse generated by a standard Tevatron synchronization CAMAC module triggers the scope's main sweep to start and digitize over either 10–20 ns of proton or antiproton bunch signal or some 1–2 μs of the electron signal. The system averages measurements over hundreds of turns in order to report positions with low error bars. The

BPM signals acquired by the digital scope are processed by a LabView program which calculates positions of all three beams. Typical statistical position measurement error is about $\approx$10–20 μm peak to peak.

The BPM pickups installed in the TEL-1 are a diagonal cut cylinder type which have shown an exceptional linearity. However, an unacceptably large 1–1.5 mm discrepancy has been observed in the reported positions of ~10 ns short proton or antiproton bunches and about 1 μs long electron pulse [48]. The major sources of the offset are thought to be the capacitance between the two plates, different stray capacitances to ground and the cross-talking between pair of electrodes which lead to significant differences of the BPM impedances for electron beam and proton beam signals, since for proton-like signal the main frequency component is about 53 MHz while for electron beam the main frequency component is less than 2 MHz.

Correspondingly, the following electron lens, TEL-2, has been equipped with a new type of the BPMs with four plates separated by grounded strips (to reduce plate-to-plate cross-talk). Together with a new signal processing algorithm which uses 5–20 MHz band-pass filtering for both electron and proton signals convoluted with Hanning window, the frequency dependent offset has been reduced to an acceptable level of less than 0.2 mm [60, 61].

In order to reduce the noise in the system, special attention has been placed on the cabling from the pick-up plates all the way to the oscilloscope. Fifty-ohm coaxial in vacuum cable is attached to each BPM plate, drawn through coaxial feedthroughs in the vacuum vessel, brought out of the Tevatron tunnel, and into the BPM electrical apparatus. With the outer conductor grounded at several places along the route (such as the vacuum feedthrough and the signal switcher) reasonable preservation of the signals has been observed.

All of these cables are also shielded in 50-Ω cabling and separated from pulsed power signals, such as the anode modulator pulses. This level of caution succeeded in preventing significant contamination of low-level signals by high-power transients.

During several years of operation, a lot of effort was put into the reduction of the electron beam imperfections and noises. For example, fluctuations in the current needed to be less than 1 %, in order to minimize the growth of 980 GeV (anti) protons emittance. Correspondingly, cathode and anode power supplies were stabilized by filtering power line harmonics and the Fermilab specific line at 15 Hz (cycle frequency of the FNAL 8 GeV Booster synchrotron). Timing jitter of the electron pulse of more than 1 ns translates into effective electron current variation as the electron pulse usually does not have a perfect flat top that results in a significant lifetime degradation of the Tevatron bunches interacting with the electron pulse. By replacing an electron pulse function generator and a delay card we managed to reduce the jitter to less than 1 ns and resolve the proton lifetime issue.

The TEL magnets have a small effect on the 980-GeV proton-beam orbit causing its distortion around the ring of about $\pm$0.2 mm. Most of the distortion comes from transverse fields in the TEL bends. Quenches of the main solenoid do not disturb the Tevatron beams significantly, but the electron beam is unable to propagate through the electron lenses. Hence, the interlock system turns off anode modulator power supplies in order to prevent electron beam generation. These power supplies are turned off if any corrector power supply, HV power supply, or vacuum gauge detects any malfunction or anomaly. The broadband impedance of the TEL components is $|Z/n| < 0.1\ \Omega$, much less than the total Tevatron impedance of $5 \pm 3\ \Omega$— and correspondingly, is of no harm to the collider beams.

A very important part of the TEL operation is the diagnostics of the Tevatron bunches themselves. Monitoring of the position, intensity, losses, emittances, betatron tunes, chromaticities of high energy bunches provided extremely useful information which allowed for optimal tune-up of the TELs. A summary of the Tevatron beam diagnostics can be found in [57].

2.3.3.1 RHIC Electron Lens Beam-Beam Overlap Monitors

The RHIC electron lenses employ a quite ingenious, fast and reliable method to monitor the overlap of the electron and proton beams. The monitor [63] is based on detection of high energy back scattered electrons which return from the interaction region all the way back to the electron gun cathode. Because of large scattering angles, the electrons come back with very large displacements making it relatively easy to detect them with a scintillator (placed near the cathode) and a photomultiplier (PMT)—see Fig. 2.34. The PMT signal varies when the x,y positions and angles of the electron beam are changed by dipole correctors. Its maximum indicates the most optimal beam-beam overlap (see alos discussion in Sect. 3.2.3).

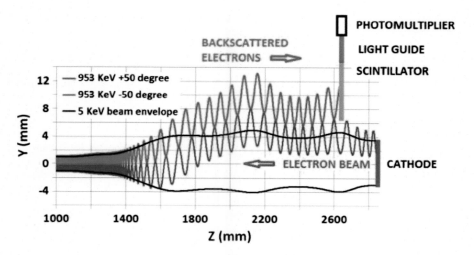

Fig. 2.34 RHIC beam overlap monitor using backscattered electrons: shown are two simulated trajectories of backscattered electrons arriving at the gun above the primary electron beam [40]

References

1. E. Tsyganov, A. Taratin, A. Zinchenko, Preprint SSCL-Report No. 519 (1993)
2. E. Tsyganov, A. Taratin, A. Zinchenko, Phys. Part. Nucl. **27**, 279 (1996)
3. V. Shiltsev, V. Danilov, D. Finley, A. Sery, Phys. Rev. ST Accel. Beams **2**(7), 071001 (1999)
4. Y. Luo, W. Fischer, Report BNL C-A/AP/286 (2007)
5. V.Shiltsev, in *Proceedings of CARE-HHH-APD LHC-LUMI-06 Workshop "Towards a Roadmap for the Upgrade of the CERN & GSI Accelerator Complex"* (16–20 October 2006, Valencia, Spain), Yellow Report CERN-2007-002 (2007)
6. Yu. Alexahin, FERMILAB-Pub-00/120-T (2000)
7. D. Shatilov, V. Shiltsev, FERMILAB-TM-2124 (2000)
8. Yu. Alexahin, D. Shatilov, V. Shiltsev, in *Proceedings of 2001 Sarantsev Seminar* (JINR, Dubna, Russia)
9. D. Shatilov, Yu. Alexahin, V. Shiltsev, in *Proceedings of 2001 IEEE PAC* (Chicago, IL, USA, 2001), p. 2002
10. Yu. Alexahin, V. Shiltsev, D. Shatilov, in *Proceedings of 2001 IEEE PAC* (Chicago, IL, USA, 2001), p. 2005
11. A. Valishev, G. Stancari, FERMILAB-TM-2571-APC (2013)
12. G. Stancari et al., Phys. Rev. Lett. **107**, 084802 (2011)
13. V. Shiltsev et al., in *Proceedings of 2008 EPAC* (Genoa, Italy, 2008), p. 292
14. G. Stancari et al., Fermilab Preprint FERMILAB-TM-2572-APC (2014), arXiv:1405.2033
15. A. Burov, G. Foster, V. Shiltsev, Fermilab Preprint FNAL-TM-2125 (2000)
16. M. Reiser, *Theory and Design of Charged Particle Beams* (Wiley, New York, 2008)
17. G.I. Budker et al., Part. Accel. **7**, 197 (1976)
18. I.N. Meshkov, Nucl. Instrum. Meth. Phys. Res. Sect. A **391.1**, 1 (1997)
19. J.R. Pierce, *Theory and Design of Electron Beams* (Van Nostrand, New York, 1954)
20. A. Sharapa, A. Grudiev, D. Myakishev, A. Shemyakin, Nucl. Instrum. Meth. Phys. Res., Sect. A **406**, 169 (1998)
21. L.P. Smith, P.L. Hartman, J. Appl. Phys. **2**, 320 (1940)
22. J.D. Lawson, *The Physics of Charged-Particles Beams*, 2nd edn. (Clarendon Press, Oxford, 1988)
23. N. Dikansky, S. Nagaitsev, V. Parkhomchuk, in Report No. FNAL-TM-1998 (1997)
24. V. Shiltsev, A. Zinchenko, Phys. Rev. ST Accel. Beams **1**, 064001 (1998)
25. G. Kuznetsov, Nucl. Instrum. Meth. Phys. Res., Sect. A **340**, 204 (1994)
26. See, e.g., I. N. Meshkov, *Transportation of Intense Beams of Charged Particles* (Budker INP, Novosibirsk, 1989) (in Russian)
27. A.V. Burov, V. Kudelainen, V. Lebedev, V. Parkhomchuk, A. Sery, V. Shiltsev, Report No. INP 89-116 1989 (in Russian); Report No. CERN/PS 93-03 (AR) (1993)
28. A.V. Burov, Report No. INP 88-124 (1988)
29. J.E. Augustin, SLAC Note No. PEP-63 (1973)
30. B.W. Montague, CERN Report No. CERN/ISR-GS/75-36 (1975)
31. A. Burov, V. Danilov, V. Shiltsev, Phys. Rev. E **59**, 3605 (1999); Report No. FNAL-Pub-98/195 (1998)
32. G. Rumolo, F. Zimmermann, Phys. Rev. ST Accel. Beams **5**,121002 (2002); see also review in Section 2.4.14 of the book A.W. Chao, et al. *Handbook of Accelerator Physics and Engineering (2nd Edition)* (World Scientific Publishing, 2013) and references therein.
33. A. Chao, *Physics of Collective Beam Instabilities in High Energy Accelerators* (Wiley, New York, 1993)
34. V.V. Danilov, E.A. Perevedentsev, Nucl. Instrum. Meth. Phys. Res., Sect. A **391**, 77 (1997)
35. G.V. Stupakov, Report No. SSCL-575 (1992)
36. D.V. Pestrikov, Nucl. Instrum. Meth. Phys. Res., Sect. A **373**, 179 (1996)
37. V. Lebedev, V. Parkhomchuk, V. Shiltsev, G. Stupakov, Part. Accel. **44**, 147 (1994)
38. B. Baklakov et al., Phys. Rev. ST Accel. Beams **1**, 031001 (1998)

39. V. Shiltsev et al., Phys. Rev. ST Accel. Beams **11**, 103501 (2008)
40. W. Fischer et al., *arxiv:1410.5315 (2014)*; also in *Proceedings of 2013 ICFA Mini-Workshop on Beam-Beam Effects in Hadron Colliders*, ed. by W. Herr, G. Papotti (CERN, March 18–22, 2013), CERN-2014-0044 (2014), pp. 109–120.
41. W. Fischer et al., in *Proc. IPAC'2014* (Dresden, Germany, 2014), p. 913
42. V. Shitsev et al., New J. Phys. **10**, 043042 (2008)
43. V. Shiltsev et al., Phys. Rev. Lett. **99**, 244801 (2007)
44. X.-L. Zhang et al., Phys. Rev. ST Accel. Beams **11**, 051002 (2008)
45. A. Ageev et al., in *Proceedings of IEEE PAC 2001* (Chicago, 2001), p. 3630
46. L. Tkachenko et al., in *Proceedings of European PAC 2002* (Paris, 2002), p. 2435
47. L. Tkachenko, Preprint IHEP 98-28 (Protvino) (1998)
48. X.-L. Zhang et al., in *Proceedings of 2003 IEEE PAC* (Portland, OR, USA, 2003), p. 1781
49. K. Bishofberger et al., in *Proceedings of 2001 IEEE PAC* (Chicago, IL, USA, 2001), p. 340
50. A. Ivanov, M. Tiunov, in *Proceedings of 2002 EPAC* (Paris, France, 2002), p. 1634
51. C. Crawford et al., in *Proceedings of 1999 IEEE PAC* (New-York, NY, USA, 1999), p. 237
52. G. Stancari et al., in *Proceedings of 2010 IPAC* (Kyoto, Japan, 2010), p. 1698
53. A.I. Pikin et al., in *Proceedings of 2011 Particle Accelerator Conference* (New York, 2011) p. 2309
54. D. Wildman et al., in *Proceedings of IEEE PAC 2001* (Chicago), p. 3726
55. Yu. Terechkine et al., Fermilab Preprint Conf-04/062 (FNAL, 2004)
56. V. Kamerdzhiev et al., in *Proceedings of 2007 IEEE PAC* (Albuquerque, NM, USA, 2007), p. 2257
57. G. Saewert, Fermilab Preprint TM-2390-AD (FNAL, 2007)
58. H. Pfeffer, G. Saewert, JINST **6**, P11003 (2011)
59. X.L. Zhang et al., in *Proceedings of 2001 IEEE PAC* (Chicago, IL, USA, 2001), p. 2305
60. V. Kamerdzhiev, in *Proceedings of 2007 IEEE PAC* (Albuquerque, NM, USA, 2007), p. 1706
61. V. Scarpine et al., in *Proceedings of 2006 Beam Instrumentation Workshop*, Fermilab, AIP Conference Proceedings 868 (AIP, Melville, 2006), ed. by T. Meyer, R. Webber, p. 481
62. R. Moore, A. Jansson, V. Shiltsev, JINST **4**, P12018 (2009)
63. P. Thieberger et al., in *Proceedings of Beam Instrumentation Workshop BIW12* (Newport News, Virginia, 2012)

Chapter 3
Electron Lenses for Beam-Beam Compensation

We begin this Chapter on the practical applications of the electron lenses for the beam-beam compensation (BBC) with experimental studies and successful experimental demonstration of the compensation long-range beam-beam effects by the electron lenses in the Tevatron collider. The head-on compensation studies in the Tevatron and successful employment of the Gaussian electron lenses for significant improvement of the RHIC luminosity are to follow. Finally, we discuss practical aspects of the beam-beam compensation with electron lenses in the LHC.

3.1 Compensation of Long-Range Beam-Beam Effects

3.1.1 Specific Requirements for Long-Range Beam-Beam Compensation in the Tevatron Collider

The beam-beam interaction between protons and antiprotons in the Tevatron took place at the two head-on interaction points located at B0 and D0 sectors, as well as at 70 parasitic crossings where the beam orbits are typically separated by about a dozen of their rms sizes. In general, the beam-beam phenomena in the Tevatron collider are characterized by a complex mixture of long-range and head-on interaction effects, record high beam-beam parameters for both protons and antiprotons (the head-on tune shifts are about $\xi^p = 0.020$ for protons and $\xi^a = 0.028$ for antiprotons, in addition to long-range tune shifts of $\Delta Q^p = 0.003$ and $\Delta Q^a = 0.006$, respectively), and remarkable differences in beam dynamics of individual bunches. Figure 1.9 - see Chapter 1 - displays the Tevatron beam tunes at the beginning of a high-luminosity HEP store on a resonance plot. Particles with up to 6σ amplitudes are presented. Small amplitude particles have tunes near the tips of the "ties" depicted for all 36 proton and 36 antiproton bunches. The most detrimental effects occur when particle tunes approach the resonances. For example, prominent

© Springer Science+Business Media New York 2016
V.D. Shiltsev, *Electron Lenses for Super-Colliders*, Particle Acceleration and Detection, DOI 10.1007/978-1-4939-3317-4_3

emittance growth of the core of the beam is observed near the fifth-order resonances (defined as $nQ_x + mQ_y = 5$, such as $Q_{x,y} = 3/5 = 0.6$) or fast halo particle loss near twelfth-order resonances (for example, $Q_{x,y} = 7/12 \approx 0.583$). Overall, the beam-beam effects at all stages of the Tevatron operation result in about 10–15 % loss in a store's integrated luminosity for a well-tuned machine but it can often be as high as 20–30 % in the case of non-optimal operation—see reviews of the beam-beam effects in the Tevatron in [1, 2].

The Tevatron beam injection and abort require three 2.6 μs gaps between 3 trains of 12 bunches each separated by 396 ns. Consequently, a threefold symmetry is expected [3] and observed [1, 3] in the pattern of the antiproton bunch orbits (which vary by about 40 μm bunch-to-bunch), tunes (which vary by up to 0.006 bunch-to-bunch) and chromaticities (6 units of $Q' = dQ/(dp/p)$ variations). Similar but smaller effects are seen in the proton bunches as well. Most profoundly these differences are seen in bunches close to the gaps, i.e., the first and the last bunch in each train. It is easy to see that these bunches encounter only one collision at the separated parasitic interaction points nearest to the main IPs, while all other bunches have them both. That one "missing" parasitic encounter is usually at the locations where the beams are very close to each other, so the effects are quite strong. Expectations were that the difference may result in very fast particle losses from the bunches at the ends of the bunch trains, so after a while these bunches vanish and then the next bunches become the "outliers", so they die in turn, and so on—therefore, the name of the phenomena—"PACMAN effect" [4]. Not only bunches at the ends are different, but in fact each bunch in each beam has a unique pattern of parasitic collisions and therefore, unique beam-beam dynamics. Betatron tunes Q_x, Q_y are thought to be one of the most important parameters tune, and they can be equalized and properly adjusted to the most optimal values by electron lenses with modulated electron currents. Indeed, a round, constant density electron beam with total current J_e, radius a_e, and interacting with, e.g., antiprotons over length L_e, will produce tune shifts of

$$dQ_z^e = -\frac{\beta_z}{2\pi} \frac{(1 + \beta_e)J_e L_e r_p}{e\beta_e \gamma_a c a_e^2}, \qquad (3.1)$$

where z stands for either x or y—that is slightly modified (1.17). If the electron beam radius a_e is several times the antiproton rms beam size, then most antiprotons have nearly equal tune shifts and the variable in time electron current can be used for the compensation of the bunch-to-bunch tune spread. Equation (3.1) shows that both horizontal and vertical tune shifts due to head-on collision with electrons have the same (negative) sign. In contrast, the long-range beam-beam proton-antiproton interaction at parasitic crossings shift horizontal and vertical tunes in opposite directions $\Delta Q_x^a = -\Delta Q_y^a$. The resulting bunch-to-bunch tune spread along the line $\Delta Q_x^a + \Delta Q_y^a$ is several times the spread along $\Delta Q_x^a - \Delta Q_y^a$ as seen in Fig. 1.9.

Obviously, two electron lenses—one at a location with the horizontal beta function larger than vertical $\beta_x \gg \beta_y$, and another one at $\beta_x \gg \beta_y$—can compensate any bunch-to-bunch tune spread and equalize both vertical and horizontal tunes simultaneously. The first one will produce a bigger tune shift in the horizontal plane, and the second in the vertical plane. That is exactly the arrangement for the two Tevatron electron lenses (TELs) at the F48 and A11 locations: the first TEL, "TEL-1," is installed in the F48 sector, where the horizontal beta-function $\beta_x = 104$ m is larger than the vertical beta-function $\beta_y = 29$ m and correspondingly, mainly affects horizontal beam tunes; while "TEL-2," the second TEL, is placed in the A11 sector, where $\beta_y = 150$ m is larger than $\beta_x = 68$ m, so that it affects the vertical tune more strongly. Note that the electron beams are round in both lenses.

Now, if one denotes the currents in the two electron lenses as $J_1(t)$ and $J_2(t)$, then the combined particles' tune shifts due to two lenses are equal to:

$$dQ_z^e(t) = \beta_{1,z} J_1(t) C_1 + \beta_{2,z} J_2(t) C_2, \quad C_{1,2} = -\frac{(1+\beta_e)L_e r_p}{2\pi e \beta_e \gamma_a c a_{1,2}^2}. \tag{3.2}$$

Full compensation of the tune spread requires the currents to be solutions of two linear equations $dQ_z^e = -\Delta Q_z^a$ (i), where i enumerates the bunch number and, therefore, $t = i \times T_B$. The required currents $J_1(t)$ and $J_2(t)$ vs the bunch number i calculated for the expected operation scenario of the Tevatron Collider Run II are shown in Fig. 3.1 [5]. The patterns of these currents have to be repeated periodically with the Tevatron revolution period of about 21 μs. The result of application of these electron lens currents ise that all particle tunes of all the bunches become identical and can be adjusted simultaneously to the most operationally advantageous betatron frequencies by usual means of the collider tune correction circuits.

Therefore, two electron lenses with beams wider than the rms (anti)proton size and approximately uniform transverse distribution of time-varied electron currents would not distort the tune footprint of each individual bunch but could compensate the bunch-to-bunch tune spread and give significant reduction of the tune area covered by the Tevatron beam, and allow for great improvement in the high energy beam dynamics (losses, lifetime, etc.). This is the aim of the long-range (or "linear",

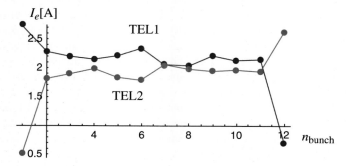

Fig. 3.1 Currents in the two electron lenses to compensate the bunch-to-bunch tune spread in the Tevatron Run II antiproton beam; see text

as the radial electron space-charge forces almost linearly depend on the radius) beam-beam compensation.

Besides that major objective, there are a number of subtle issues relevant to compensation of the bunch-to-bunch tune spread which have been studied numerically with beam-beam simulation code LIFETRAC prior to the practical implementation of the method, e.g., the optimal electron beam current distribution and size, straightness of the electron beam, misalignment of the electron and (anti) proton beams, the time to bring the electron and (anti)proton beams into collision, the effects of noise in the electron beam, and the effect on the other (proton) beam [6]. The major results of those tracking studies work were:

1. Significant positive effect of the tune shift compensation by the electron lenses was confirmed in simulations—see Fig. 3.2 from [6];
2. The minimum acceptable ratio of the electron beam radius to the rms antiproton beam size needs to be about 3 to avoid growth of large betatron amplitude tails in the transverse plane;
3. The acceptable misalignment of the electron and antiproton beams depends on the e-beam radius and was found to be in the range of 0.1–0.5 mm;
4. The antiproton emittance growth due to electron beam noises was found to be in good agreement with the theoretical predictions, see Chap. 2 above;
5. The separation between e- and p-beams at proposed TEL locations is quite acceptable, so the expected effects of the TEL on the proton beam are small, while electrons are centered on the nearby antiprotons.

3.1.2 Initial Experimental Characterizations of the Electron Lens Effects

A series of beam experiments were performed with the electron lenses to characterize dependencies of the particle betatron tunes on electron beam current, energy and position, and to evaluate effects arising from imperfections and noises of the electron beam, as well as the effect of the electron current-density profile on the Tevatron beam losses. The shape of the electron current density distribution is determined by the electron gun (geometry of its electrodes and voltages). In the beam studies, three types of electron guns generating flat-top, Gaussian, and smoothed-edge-flat-top (SEFT) distributions were employed, as shown in Fig. 2.19. Below we will specifically mention which gun was being used for each of the experimental studies. Figure 3.3 schematically depicts all three beams— electron (green), proton (blue) and antiproton (red)—inside the A11 TEL beam tube in the configuration when the electrons are set on the protons. Relative sizes of 1, 2, and 3σ ellipses for the 980-GeV protons and antiprotons are indicated as well as the size of the round (SEFT) electron beam.

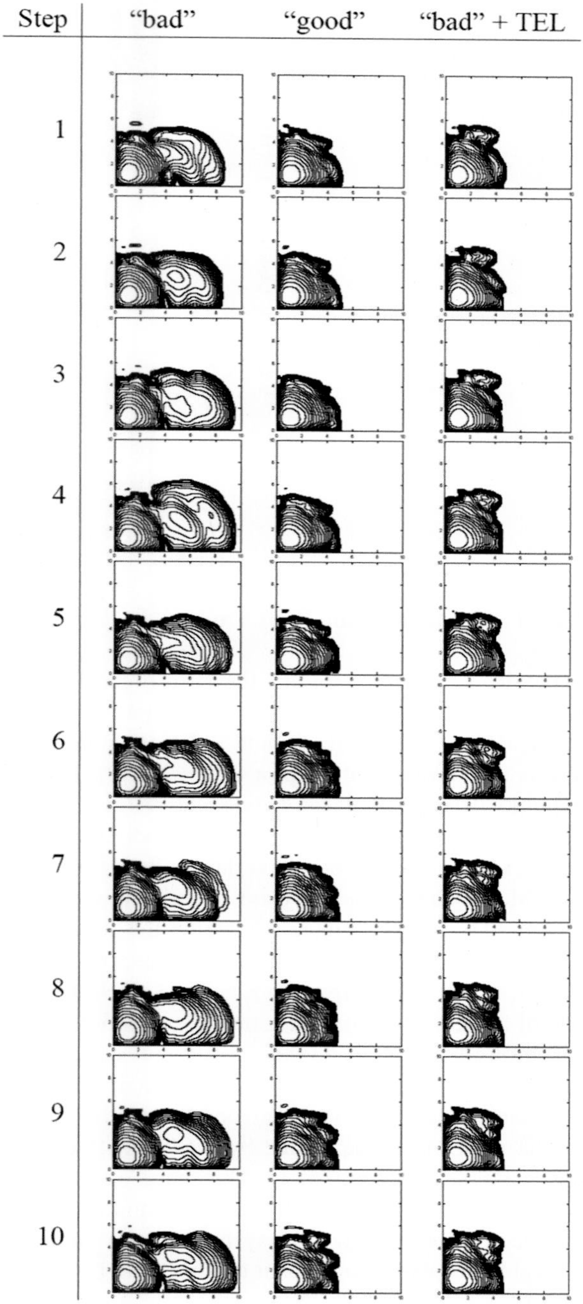

Fig. 3.2 Dynamics of the distribution of the antiprotons in the plane of normalized betatron amplitudes A_x and A_y. Each row represents a consecutive step in the numerical tracking (ten steps of the 300,000 turns of beam-beam tracking, each step corresponds to about 6 s of beam time in the Tevatron). It is clearly seen how the large-amplitude tails grow in the "bad" (not optimal) working point Q_x,Q_y (*left column*), compared to an "optimal" one (*center column*) - both cases without electron lenses - and how the TEL-induced tune shifts toward the "optimal" working point Q_x,Q_y improves the situation significantly. The distance between successive contour lines is sqrt(e)

Fig. 3.3 Schematic of the transverse electron beam alignment with respect to the proton and antiproton beams [7]

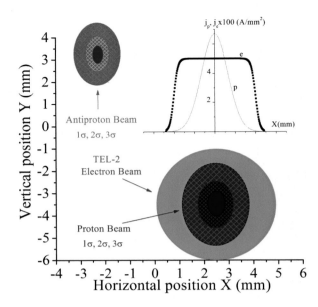

3.1.2.1 Betatron Tune-Shift Studies

In the first series of the studies, the shifts of the Tevatron betatron tunes by the TELs were measured in great detail and compared with theory (3.1).

21 MHz and 1.7 GHz Schottky detectors are used in Tevatron to measure the tunes of the proton and antiproton bunches [8]. The 21 MHz vertical and horizontal Schottky detectors have very good tune resolution but are not directional and do not allow for measurement of the tunes of individual bunches. The 1.7 GHz detectors are less accurate but they have larger bandwidth, allowing for measurements of individual bunches and they are directional, so, proton and antiproton signals are measured separately. Figure 3.4 depicts the 21 MHz Schottky spectra during one of the TEL-1 experimental studies: only three proton bunches were circulating in the Tevatron (without antiprotons), and the electron current pulse was timed on only one of the three bunches. A series of synchro-betatron sidebands, separated by the synchrotron tune $Q_s \approx 0.0007$, on the left correspond to the signal from the two bunches not affected by the TEL-1, and their central line (highlighted by a marker) was found at the fractional horizontal tune of $Q_x = 0.5795$. A similar series of lines on the right, which generated by the TEL-affected proton bunch, was shifted by $dQ_{x,} = 0.0082$ to $Q_x = 0.5877$. The shape of the Schottky spectra depends on the proton intensity, machine chromaticity, tuning, working point, etc. Application of the electron beam may or may not cause a variation of the spectral shape. The typical tune measurement error for the 21 MHz Schottky detector is estimated to be about $\delta Q \approx \pm 0.0002$.

After proper synchronization of the electron pulse for the maximum effect, the dependence of dQ_x on the peak electron current was studied [7]—see the results in

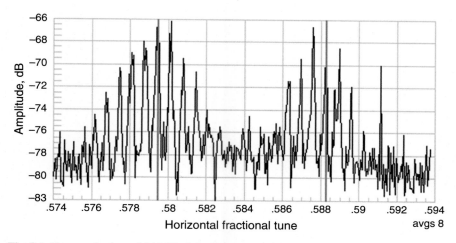

Fig. 3.4 Tevatron horizontal 21-MHz Schottky spectrum of three 980 GeV proton bunches with one of them tune-shifted to the right by TEL1 (electron current $J_e = 2.6$ A, electron energy $U_c = 7.6$ kV, flat-top electron gun) [7]

Fig. 3.5 - and compared with the tune shift theory (3.1). The theoretical dependence is nonlinear because the electron energy inside the vacuum pipe (and thus β_e) decreases with increasing current due to the electron space charge, $U_e = U_{c.} - gQ_{SC}$, where g is a factor depending on the chamber and beam geometries. In Fig. 3.5a, the tune shift data measured by 21 MHz Schottky detector during four selected experiments with TEL-1 is shown. The TEL cathode voltage was set to -7.5 kV in experiments #4 and #14, to -8 kV in the experiment #18 and to -4.7 kV in the experiment #25. The voltage difference is the reason for significant variation of slopes in Fig. 3.5a. The experiment labeled as Experiment #4 is in fact one of the earliest with the TEL-1 after it was installed in the Tevatron and the large discrepancy between measured tune shift and theoretical prediction was due to poor electron beam alignment (see also discussion below). In the other three experiments with better alignment, the maximum discrepancy did not exceed ~20 % at $J_{e.} = 2$ A. There are systematic errors in a number of the parameters used in numerical calculating (3.1). For example, a_e^2 is known only to within ± 10 %, the effective length L_e depends on the precision of the steering and may vary within ± 10 %, and the electron current calibration contributes about ± 5 % error [9]. Figure 3.5b presents the vertical tune shift induced by the TEL-2 electron current from the SEFT gun. There is an excellent agreement between the tune shift measured by the 1.7 GHz Schottky tune monitor and the theory. The dependence of the tune shift on the electron energy also agrees with the theoretical predictions; pertinent results displayed in Fig. 3.6 show 980-GeV antiproton tune-shift measurements at various cathode voltages U_c, ranging from -6 to -13 kV. As the total electron beam current (which is determined by the gun cathode–anode voltage difference, and shown by the dashed line) was kept constant, the total electron space-charge Q_{SC} grew for smaller values of U_c, inducing correspondingly larger tune shift.

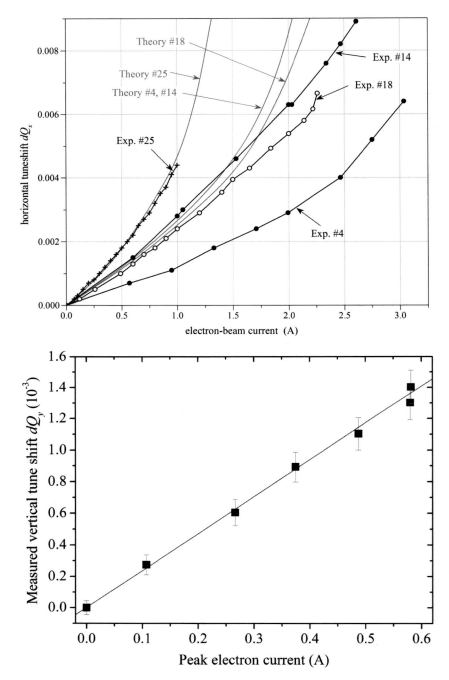

Fig. 3.5 Shift of the 980-GeV proton betatron tunes by electron lenses : (**a**) *Top*—horizontal tune shift vs. the electron current in TEL-1, over several separate experiments (*flat-top* beam profile); (**b**) *bottom*—vertical tune shift vs. TEL-2 current (SEFT gun). Solid lines are theoretical predictions according to (3.1) [7]

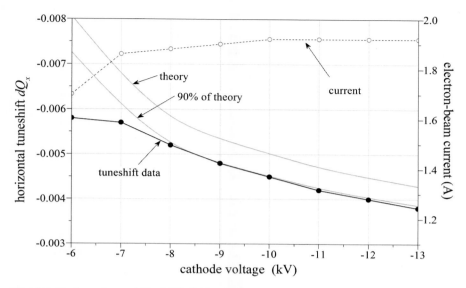

Fig. 3.6 Horizontal tune shift of 980-GeV antiprotons versus cathode voltage (electron energy). TEL-1 with the flat-top electron gun was used [7]

Figure 3.7 illustrates how the proton tune shifts depend on the time delay between the 2 A electron pulse and the arrival of the proton bunch. With a small correction for the electron beam propagation time along the TEL interaction region (~50 ns), the tune shift basically follows the electron pulse shape. One important conclusion is that the electron pulse is short enough to allow shifting the tune of any given bunch without touching its neighbor's 396-ns away. This feature was extremely useful over the entire series of beam studies with the TELs, as it allowed us to vary electron beam parameters and to tune up the lens affecting just one bunch out of 36 circulating in the machine. As seen in Fig. 3.7, the horizontal tune shift is about four times the vertical one $dQ^e_x/dQ^e_y = 0.0037/0.0008 = 4.6$, which is close to the design beta function ratio $\beta_x/\beta_y = 101/28 = 3.6$. The remaining discrepancy can be explained by the beta-function measurement error, which could be as big as $\pm 20\%$, or a small ellipticity of the electron beam, or a mis-steering of the electron beam, which may play a role if it is not smaller when compared with a_e.

As long as the proton beam travels inside a wider electron beam, the proton tune shift does not depend much on the electron beam position d_x, d_y; for example, in the case of a 1-A electron beam, $dQ^e_x(d_x, d_y) \approx dQ_{max} = 0.0021$ if the displacement $|d_{x,y}| < 2$ mm, as illustrated in Fig. 3.8. However, when the distance between the centers of the two beams exceed the electron-beam radius a_e then one expects $dQ^e_x(d_x, d_y = 0) \approx -dQ_{max}/(d_x/a_e,)^2$, $|d_x| > a_e$, and $dQ^e_x(d_x = 0, d_y) \approx +dQ_{max}/(d_y/a_e,)^2$ $|d_y| > a_e$. Such a change of the sign of the tune shift is clearly seen in Fig. 3.8.

To summarize, the experimentally observed shifts of the betatron tunes of 980 GeV protons and antiprotons due to TELs agree reasonably well with theoretical predictions (3.1).

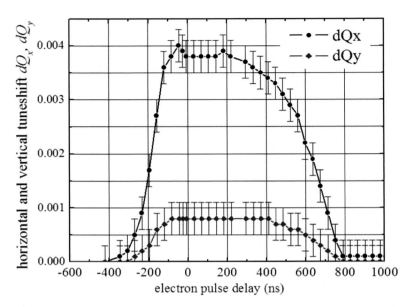

Fig. 3.7 TEL-1-induced shift of 980 GeV proton horizontal (*black line*) and vertical (*blue line*) betatron tune versus delay time for an 800-ns long pulse of electrons (1.96 A peak current, −6.0 kV cathode voltage, flat-top gun) [7]

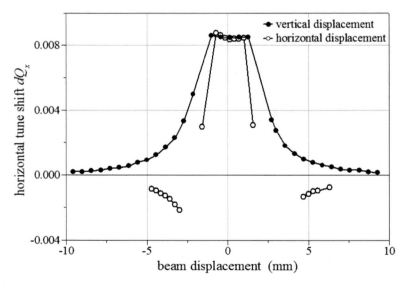

Fig. 3.8 TEL1-induced shift of 980 GeV proton horizontal betatron tune versus vertical (*filled circles*) and horizontal (*open circles*) electron beam displacement ($J_e = 1$ A, $U_c = 6.0$ kV, flat-top electron gun) [7]

3.1.2.2   Studies of Electron Beam Fluctuations Effects

Fluctuations in the electron beam can lead to several phenomena in the high energy beams: (a) turn-by-turn electron current jitter and transverse electron beam position fluctuations can blow up the transverse (anti)proton emittance—e.g., theory presented in Chap. 2 above predicts sizable growth if the rms current fluctuations δJ exceed (3–10)mA or if the position jitter δX in a multi-Ampere beam is bigger than 0.2 μm; (b) electron pulse timing jitter results in similar effects if the electron current pulse does not have a flat-top; (c) low-frequency variations of the parameters may result in the orbit or/and tune variations leading to a faster dynamical diffusion. High-frequency current fluctuations measured directly from the TEL-1 BPM signals using a 15 bit ADC segmented memory scope showed $(\delta J/J) \sim$ $(4$–$10) \times 10^{-4}$ for pulses of current $J \approx 0.3$–0.5 A; one could also estimate an upper limit on the electron beam position stability of about 10 μm [10].

To observe the effect of the TEL current fluctuations on the antiproton emittance growth, the electron gun HV modulator pulse circuit was modified to produce a random-amplitude pulse. This is established by setting an average pulse amplitude modulated by a noise generator. At different noise levels, the 980 GeV antiproton bunch emittance is observed long enough to record its growth by so called "Flying Wires" beam size monitors [8]. Figure 3.9 shows that the emittance growth increases—as expected from the theory—with the square of the amplitude fluctuations. The Tevatron high-intensity proton and antiproton beams, without the TEL, have a typical emittance growth of 0.04–0.2π mm-mrad/h. If the TEL is allowed to

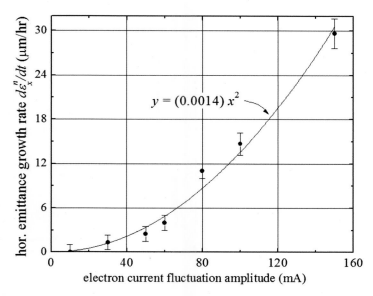

Fig. 3.9 Horizontal emittance growth rate of 980 GeV antiprotons vs TEL-1 electron current fluctuation amplitude (flat-top electron gun) [7]

only enhance the emittance growth by 0.01 mm-mrad/h, added in quadrature to the Tevatron's inherent emittance growth, then according to the measured dependence in Fig. 3.9 this limit corresponds to about 3 mA peak-to-peak current variation.

Another source of fluctuations is timing jitter. It was noted that a large timing jitter of about 10 ns peak-to-peak (due to an instability of the synchronization electronics) leads to a detectable emittance growth and to a significant increase in the 21 MHz Schottky detector signal power. This effect was particularly large on the rising and falling slopes of the electron pulse (where the derivative of the electron current dJ_e/dt is large). Elimination of the source of the instability and use of optical cables for synchronization of the TEL pulsing with respect to the Tevatron RF allowed us to reduce the jitter to less than 1 ns and to bring the corresponding emittance growth to within the tolerable level. In nominal operation conditions of the TEL, without the noise generator, with low-frequency current variations reduced to under 5 mA, and with the timing jitter under control, we observed no detectable additional emittance growth within the resolution of our beam size monitors $(d\varepsilon/dt)_{rms} \sim 0.02\pi$ μm/h.

Monitoring the 21 MHz Schottky power is useful for studying the effect of the beam displacement. Figure 3.10 presents the dependence of the Schottky power on the TEL2 vertical beam position (the electron beam is perfectly aligned with the proton beam horizontally). An additional electron current noise of about 50 mA peak-to-peak was induced in order to make the effect more prominent. The study was performed at the end of the Tevatron HEP store no. 5152 with both proton and antiproton beams present. One can see that the Schottky power rises with increasing

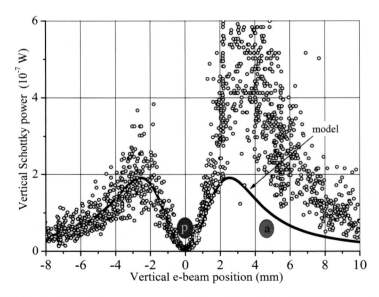

Fig. 3.10 Vertical Schottky power vs. vertical position of the TEL-2 electron beam in HEP store no. 5152. Electron current noise amplitude about $\delta J_e = 50$ mA peak-to-peak (SEFT electron gun) [7]

separation of the electron and proton beams—approximately as $P \sim E_y^2$, where the electric field due to the electron space charge is given by (1.16). The asymmetry of the measured power with respect to vertical position can be explained by the effect of the TEL2 on the antiproton beam—its position is indicated by the red ellipse in Fig. 3.10; see also Fig. 3.3 for reference (note that the 21 MHz Schottky monitor is not directional and reports the transverse Schottky signal power induced by both the proton and antiproton beams circulating in the machine).

"Tickling" (intentional transverse excitation) of the proton orbit by the electron beam can be used for electron beam steering. The idea is similar to the "K-modulation" in the beam based alignment [11]: variation of the electron current in the electron lens causes variations in the proton beam orbit around the ring if the electron lens beam is not centered. Figure 3.11 shows the rms amplitude of the vertical proton orbit variation at the Tevatron BPM located in the A0 sector vs. the vertical displacement of the TEL-1 electron beam at the F48 location. The current of the latter was harmonically modulated at 107 Hz as J_e [A] $= 1.02 + 0.18 \sin$ $(2\pi t*107$ Hz). The amplitude becomes equal to 0 if the proton beam goes through the center of the electron beam. Maximum amplitude of the orbit response at 107 Hz is about a few micrometers. The 7-mm distance between the two peaks reflects an effective diameter of the electron current distribution, and thus, indicates some angular misalignment of the electron beam because it exceeds the electron beam diameter $2a_e \approx 3.5$ mm. Therefore, steering by the orbit tickling should aim not only on the search of the minimum orbit response, but also on getting the two maxima closer to each other.

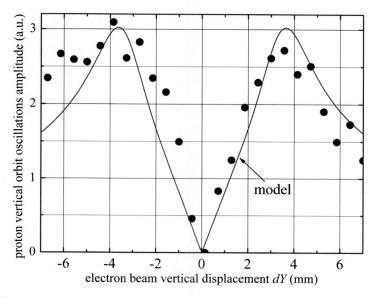

Fig. 3.11 K-modulation position scan with TEL-1 (flat-top electron gun) [7]

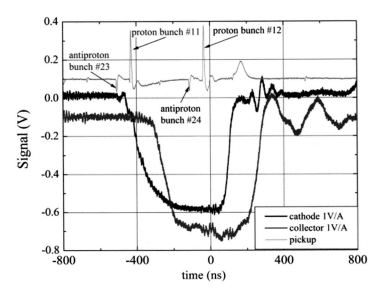

Fig. 3.12 Longitudinal waves in the TEL2 electron beam excited by the interaction with a proton bunch: *red*—TEL2 beam pickup signal, *black*—electron current measured at the SEFT electron gun cathode, *blue*—electron current measured at the collector [7]

Yet another indication of a good steering of the electron beam is the observation of longitudinal space-charge waves in the electron beam induced by the proton bunches. Figure 3.12 presents digital scope records of the TEL-2 BPM signal (pickup) and electron current pulses measured at the cathode and at the collector of the lens. The difference between the last two is the electron current pulse in the collector exhibits additional waves (wiggles starting around $t = 0$ ns) due to interaction with the protons. The amplitude of the waves is about 5 % of the maximum total electron current. Any significant separation of the electrons and protons (several mm transversely, or timing the electron pulse away from the proton bunches, e.g., in the abort gap) leads to disappearance of the waves and to the collector signal becoming like the one at the gun. Detailed theoretical analyses and extensive experimental studies of these waves are reported in [12].

3.1.2.3 Effect of Electron Beam Current Profile on High Energy Beam Lifetime

Typically the lifetime of (anti)proton bunches in the Tevatron, before they are brought to collisions with protons and without the TEL, is on the order of 200–600 h. As a bunch traverses the Tevatron ring billions of times, several mechanisms contribute to a gradual growth in its emittance of about 0.04–0.2π mm-mrad/h. These include: residual-gas scattering, intra-beam scattering, and fluctuations in the ring elements [2]. As the bunch size increases, particles gradually diffuse to larger oscillation amplitudes until they finally reach some aperture restriction, usually one of many retractable collimators inserted in the Tevatron beam pipe. As shown in [1] and as will be discussed further below, the beam-beam interaction with either the electron beam or the opposite high-energy beam can lead to a significant decrease of the beam lifetime.

Originally, it was planned to generate an electron beam wide enough to cover all of the high energy proton or antiproton beams—and this large size was thought to be helpful to maintain low particles loss rates. However in reality there are always particles with amplitudes beyond the electron beam cross section. For such particles with oscillations larger than the size of the electron beam, the electric field due to the electron space charge is no longer linear with the transverse displacement and the resulting nonlinearities may significantly change the particle dynamics depending on the electron current distribution. As it was found experimentally, in the worst case of the flat-top electron beam, steep electron beam edges act as a "soft" collimator, since the outlying particles are slowly driven out of the bunch until they eventually hit the collimators.

A convenient way to measure this effect is to observe the bunch size as the TEL trims away extraneous particles. In Fig. 3.13, one bunch was monitored over one hundred of minutes as the TEL-1 was "shaving" (removing the outer transverse region) the bunch size. The current of the TEL was initially set to 1 A for the first 45 min. After a 10 min respite, the current was increased to 2 A (these settings are shown above the plot). After about 85 min, the TEL-1 was purposefully mis-steered in order to observe a "blowup" in the bunch sizes. The upper data in Fig. 3.13 show

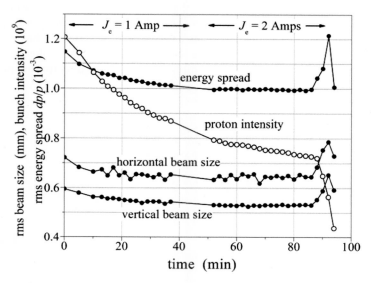

Fig. 3.13 Scraping of a proton bunch due to interaction with the TEL-1 electron beam (*flat-top* electron current distribution) [7]

the horizontal and vertical beam sizes measured many times during this process. Also indicated is the longitudinal bunch size.

The open circles show the intensity of the bunch during this process. One can see a fast initial decreasing of sizes, but after about 10 min, the rate of decrease drops significantly; this implies the large-amplitude particles have been removed, and the core is more stable inside the electron beam. In addition, the increase of the TEL-1 current to 2 A was expected to worsen the bunch-size lifetime, but the smaller bunch was well preserved for the remaining time that the TEL-1 electron beam was on and centered on the proton beam. The stability of the bunch size is remarkable, suggesting that the flattop profile was ideal for the small bunch size.

The bunch intensity decay rate also decreases significantly after a short interval of faster losses, and when the electron current is doubled, the decay rate is nearly unchanged. After the bunch was observed for a while, the electron beam was moved transversely so that the bunch intercepted the edge of the electron beam. As expected, the particles were suddenly experiencing extremely nonlinear forces, causing emittance (and size) growth, shown by the bump in the upper plot of Fig. 3.13, and heavy losses, shown by the fast decline of the lower plot.

Figure 3.14 presents the results of an experiment where the proton loss rate, as indicated by the CDF detector beam-loss monitors, was measured as the TEL current was changed. This test was performed with the flat-top electron beam centered on a single proton bunch, and the cathode voltage was set at -10 kV. The losses varied from about 250 Hz at low currents to 1 kHz at the highest currents. At zero current, the average loss rate was approximately 230 Hz over a large portion of an hour. The losses data were converted into lifetimes, producing more tangible results. This conversion was straightforward, since the lifetime is given by $\tau = -kN/(dN/dt)$, where N was the current total number of particles and (dN/dt) was the loss rate measured by the beam-loss monitors. The constant k was determined from a calibration test. During a period of a couple hours, the bunch was allowed to proceed without any changes made to the TEL-1. After this amount of time, the number of particles had diminished enough to directly compute the

Fig. 3.14 Dependence of the proton bunch intensity lifetime on the TEL-1 current (*flat-top* electron current distribution) [7]

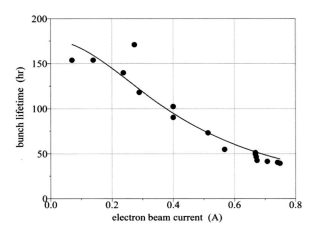

lifetime. Since the number of particles and the average loss rate were also known, the constant k could be derived for the conditions of the experiment [13].

The maximum current in this experiment was about $J_e = 0.75$ A and the corresponding proton tune shift was about $dQ_x = 0.0022$. Despite of the small tune shift, the proton bunch lifetime at the higher electron-beam currents was less than 50 h, significantly less than the typical 175-h lifetime without interference. While it was impossible to guarantee that the electron beam was perfectly centered on the proton orbit, adjustments of the beam position yielded no improvement in the bunch lifetime. The solid line in Fig. 3.14 represents the fit τ^{-1} [1/h] $= 1/150 + J_e^2/30$. In this experiment, the electron radius was $a_e = 1.6$ mm, and the proton rms beam size at the location of the TEL-1 was about $\sigma_x = 0.8$ mm, corresponding to an rms normalized emittance of about 5π mm-mrad.

The examples presented above and the unsatisfactory low beam lifetime in the early beam-beam compensation experiments during HEP stores, convinced us that sharp beam edges of the flat-top electron distribution introduce severe constraints on the performance of the TEL-1. A Gaussian gun was designed to obtain much smoother edges, so that particles at large betatron amplitudes would not feel strong nonlinear space-charge forces. Figure 2.19 depicts the current density profiles for the flat-top and Gaussian gun. In order to quantify the differences between these two guns, a scan of working points (Q_x, Q_y) was performed with each of them. In this test, the Tevatron horizontal and vertical tunes were independently adjusted to cover approximately a 0.020 span in both Q_x and Q_y. By adjusting the betatron tunes in 0.002 increments, the loss rate was measured, recorded, converted to a lifetime, and plotted in Fig. 3.15.

To simplify the interpretation of the results, both guns were set to currents such that the horizontal tune shift was $dQ^e_x = 0.004$ and the vertical was $dQ^e_y = 0.0013$. The Tevatron is equipped with tune-adjustment quadrupole magnet circuits, which provided a convenient way to adjust the tune. Confirmation of the correct tune was possible with the Schottky tune monitors, but sometimes when the particle loss rate was too high, an accurate measurement of the tune was difficult to determine. Whenever the tunes were adjusted, a short amount of time was needed before the loss rate stabilized. Sometimes it reacted quickly, while at other times it required a longer period before a specific loss rate could be determined. The number of protons in the test bunch was measured throughout the experiment period, and a calibration test was performed as described above. That allowed the loss-rate data to be converted into lifetimes as shown in Fig. 3.15.

The shaded scale shown on the right side of the scans indicates the lifetime in hours witnessed at each data point. In order to more effectively convey the regions of high and low lifetime, a two-dimensional interpolation algorithm turned the individual data points into a smooth, shaded surface. Contour lines are drawn at multiples of 20 h. Unfortunately, the regions covered by the two scans do not span exactly the same tune space, but there was a sufficient overlap to make the significant differences between the flattop and the Gaussian guns apparent. In Fig. 3.15a, the flat-top gun usually produced poor lifetimes. This implies that the TEL-1 flattop gun tended to excite oscillations in at least some portion of the bunch

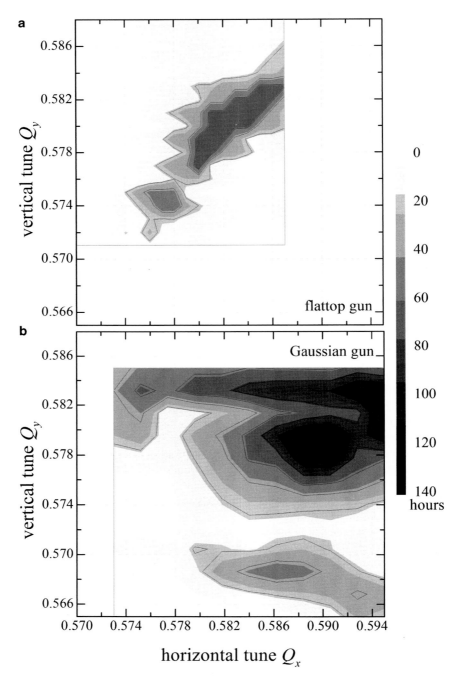

Fig. 3.15 Contour plots of proton bunch lifetime scans over a range of vertical and horizontal betatron tunes : (**a**) *top*—with $dQ^e_x = 0.004$ induced by TEL-1 with the flat-top electron gun; (**b**) *bottom*—with $dQ^e_x = 0.004$ induced by TEL-1 with the Gaussian electron gun [7]

particles, and the recorded lifetimes were usually less than 10 h. However, in the tune space region near the main diagonal $Q_x = Q_y$, there is a relatively consistent pattern of lower losses. Along this strip, lifetimes as high as 70 h were observed, almost as high as the lifetime of the bunch unaffected by the TEL-1.

The large regions of low lifetime again support the hypothesis that the flattop electron beam with sharp edges is adversely affecting protons. The outlying particles, witnessing strongly nonlinear focusing forces from the electron beam, do not survive as long as the core particles. Through the majority of the tested tune space, these particles are lost very quickly, and the gradual emittance growth of the core protons constantly feeds these losses. Only in a small working-point region do the outlying particles not escape so nimbly, slowing the rate at which particles are lost.

The second scan, in Fig. 3.15b, shows the massive difference the Gaussian gun had on the lifetime. The highest measured lifetimes were around 130 h, almost indistinguishable from the bunch lifetime without the TEL-1. Much larger regions of lifetimes over 20 h can also be seen. The fact that the highest lifetimes are nearly the same as for the unperturbed proton bunch bolsters the idea that TEL-1 fluctuations cannot by themselves, remove particles from the bunch completely. Instead, it is believed that the fluctuations contribute to a gradual emittance growth, and because there are no strongly nonlinear edges to the electron beam, the protons are still stable at larger orbits. This interpretation explains why a much larger percentage of the tested tune space offered moderate lifetimes than for the flat-top gun and why the best lifetimes observed are significantly longer with the Gaussian gun.

A significant improvement of the particle lifetime thanks to the employment of the Gaussian electron gun was critical for the first observation of a successful beam-beam compensation of the antiproton emittance growth (see Sect. 3.2.2 below).

3.1.3 Successful Compensation of the Long-Range Beam-Beam Effects in the Tevatron Collider: Improvement of Proton Beam Intensity and Luminosity Lifetimes by Electron Lenses

In 2004–2006, as an outcome of the initial TEL studies described above, four very important changes to the Tevatron and TEL operation were introduced and these changes allowed for a very regular, repeatable and successful employment of electron lenses for beam-beam compensation (BBC). Firstly, the Tevatron automatic orbit stabilization system was installed and commissioned [14], so that typical store-to-store orbit changes as well as low-frequency orbit drifts during HEP stores do not exceed 0.1 mm (high frequency orbit jitter is still uncorrected, but it is not very significant—about 0.02–0.04 mm peak-to-peak). Secondly, a new signal processing technique was introduced for the TEL BPMs which reduced the frequency dependence of the monitors from 0.5 to 1.5 mm down to about 0.1 mm [15]. Thirdly, the second electron lens was built and installed at the A11 location of the Tevatron ring which allowed dedicated beam-beam compensation studies very often—ultimately, in every HEP store—with one of the lenses while the other lens was always dedicated to the operational abort gap cleaning (a standard TEL function since early Run II—see Chap. 4.2 below and [16]). Last, but not least, electron guns with "smoothed-edge-flat-top" current distributions were designed, built and installed in both TEL-1 and TEL-2. All the results presented in this Chapter are obtained with the SEFT electron guns. After the commissioning of all these features and attainment of stable operation, a significant improvement of proton beam lifetime under the action of electron lenses was demonstrated [17].

A significant attrition rate of the protons due to their interaction with the antiproton bunches, both in the main IPs and in the numerous long-range interaction regions is one of the most detrimental effects of the beam-beam interaction in the Tevatron [1]. This effect is especially large at the beginning of the HEP stores where the total proton beam-beam tune shift induced by the antiprotons at the two main IPs (B0 and D0) can reach the values of $\xi^{proton} = +0.020$. Figure 3.16a shows a typical distribution of proton loss rates at the beginning of a high-luminosity HEP store. Bunches numbered 12, 24, and 36 at the end of each bunch train typically lose about 9 % of their intensity per hour while other bunches lose only 4–6 % per hour. These losses are a very significant part of the total luminosity decay rate of about 20 % per hour (again, at the beginning of the high luminosity HEP stores). The losses due to inelastic proton-antiproton interactions $dN_p/dt = \sigma_{int} L$ at the two main IPs ($\sigma_{int} = 0.07$ b) are small (1–1.5 %/h) compared to the total losses. Losses due to inelastic interaction with the residual vacuum are less than 0.3 %/h. The single largest source of proton losses is the beam-beam interaction with the antiprotons. Such conclusion is also supported by Fig. 3.16a, which shows a large bunch-to-bunch variation in the proton loss rates within each bunch train, but very similar rates for equivalent bunches, e.g. bunches numbered 12, 24, and 36. On the contrary, antiproton intensity losses dN_a/dt are about the same for all the bunches—see Fig. 3.16b—as they are mostly due to luminosity burn-up and not determined by beam-beam effects.

The remarkable distribution of the proton losses seen in Fig. 3.16, e.g., particularly high loss rates for bunches numbered 12, 24, 36, is usually thought to be linked to the distribution of betatron frequencies along the bunch trains bunch. Bunches at the end of the trains have their vertical tunes closer to the $7/12 \approx 0.583$ resonance lines—see Fig. 3.17—and therefore, the higher losses. The average Tevatron proton tune Q_y of about 0.588–0.589 lies just above this resonance, and the bunches at the end of each train, whose vertical tunes are lower by $\Delta Q_y = -(0.002–0.003)$ due to the unique pattern of long-range interactions, are subject to stronger beam-beam effects. The tunes Q_y, Q_x are carefully optimized by the operation crew to minimize the overall losses of intensity and luminosity. For example, an increase of the average vertical tune by quadrupole correctors is not possible because it usually results in higher losses and "scallops" as small amplitude particle tunes move dangerously close to the $3/5 = 0.600$ resonance (see Fig. 1.8).

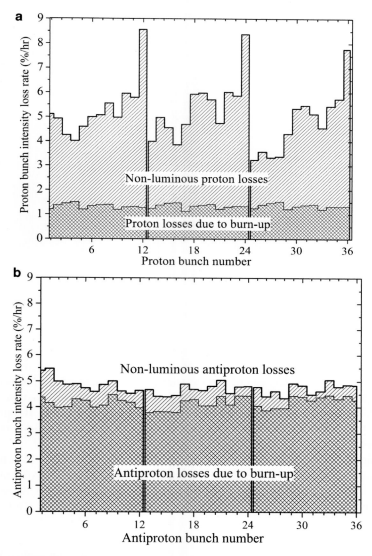

Fig. 3.16 (**a**) *Top*—proton-bunch intensity loss rates and (**b**) *bottom*—antiproton-bunch intensity loss rates at the beginning of the Tevatron store no. 5155, Dec. 30, 2006, with an initial luminosity $L = 250 \times 10^{30}$ cm^{-2} s^{-1} [7, 17]

When properly aligned, the TEL-2 electron beam focuses protons and thus, produces a positive vertical tune shift of the proton bunch it acts on, proportional to the electron current—as depicted in Fig. 3.5—and, therefore it should reduce the losses. A preliminary alignment of the electron beam has been done by relying on the TEL beam position measurement system. However, additional fine tuning is

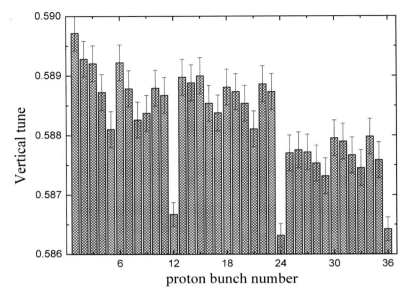

Fig. 3.17 Proton bunch tunes measured by Digital Tune Monitor [18] at the beginning of store no. 5301 (March 3, 2007) with an initial luminosity $L = 203 \times 10^{30}$ cm^{-2} s^{-1} [7]

usually necessary to achieve the best possible compensation. Measurements of the proton loss rate versus the electron beam position with an increased electron current were performed at the very end of a store, when no beam-beam related losses occur. This approach allowed us to determine the optimal electron beam position. Since the Tevatron orbit is kept stable by the orbit feedback system within 100 μm, the end-of-store values can be used throughout other stores, unless an optics change is introduced.

In one of the very first BBC demonstration experiments [17], the TEL-2 electron pulse was timed onto bunch P12 (proton bunch number 12) without affecting the other bunches. Figure 3.18a shows that when the TEL peak current was increased to 0.3 A, the lifetime $\tau = N_p/(dN_p/dt)$ of bunch P12 went up to 17.4 ± 0.1 h from an initial 8.75 ± 0.1 h (in other words, the loss rate $1/\tau$ improved twofold from about 11.4 %/h to some 5.7 %/h). At the same time, the lifetime of bunch P24, an equivalent bunch in another bunch train, remained low and did not change significantly ($\tau = 8.66$ h lifetime slightly improved to 10 h due to natural reasons—see discussion below). The TEL was left on P12 for the first 1.5 h of the store and the intensity decay of that particular bunch was one of the lowest among all 36 proton bunches—as shown in Fig. 3.18b which presents the loss rates corrected for luminous intensity decay which is not due to beam-beam effects $(dN_p/N_p)_{NL}/dt = (dN_p/N_p)_{total}/dt - \sigma_{int} L/N_p$. It is noteworthy, that the vertical tune shift caused by such a moderate electron current $J_e = 0.3$ A is about $dQ^e_y = +0.0007$ (as can be seen in Fig. 3.5b) and it is not sufficient for P12 to reach the average tune $dQ^e_y < |\Delta Q_y| \approx 0.002$. Therefore, the TEL-induced tune shift could not be

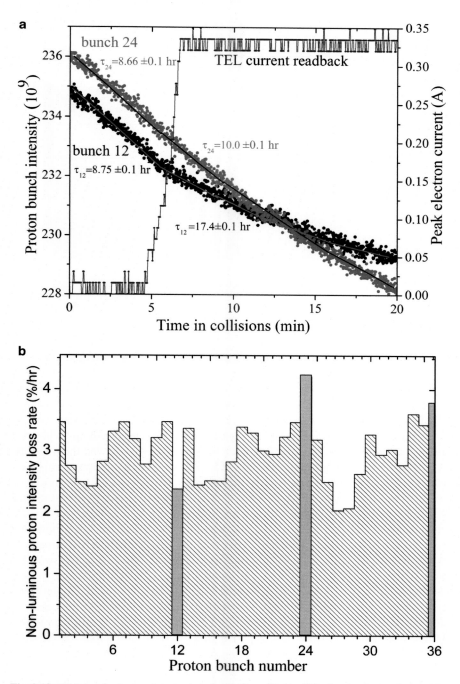

Fig. 3.18 (**a**) Intensity decay of proton bunch 12 affected by the TEL-2 and reference bunch 24 at the beginning of store no. 5123 with an initial luminosity $L = 197 \times 10^{30}$ cm^{-2} s^{-1}. The blue line shows the measured TEL-2 peak current; (**b**) the average non-luminous bunch intensity loss rate in the first 1.5 h of the store [7, 17]

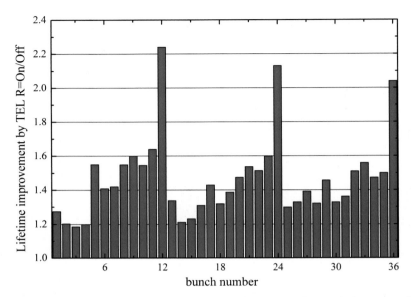

Fig. 3.19 Proton bunch lifetime improvement due to TEL-2 (DC regime) early in store no. 5183 with initial luminosity $L = 253 \times 10^{30}$ cm^{-2} s^{-1} [7]

considered as the only mechanism responsible for the significant lifetime improvement in that experiment.

To explore more detailed electron lens effects, another series of BBC studies was performed in one of the highest luminosity Tevatron stores no. 5183, in which the TEL-2 operated in a DC regime with $J_e = 0.3$ A—providing the same effect on all proton bunches in the beam—and has been regularly turned off and on. When the TEL-2 was turned on at the very beginning of the store, it improved the intensity lifetime of all the bunches, as presented in Fig. 3.19, although the largest improvement $R = 2.2$, defined as the ratio of the proton lifetime with the TEL and without it, was observed for bunches P12, P24 and P36, as expected. Later in the store, the TEL-2 was turned off for some 20 min and then, by use of magnet correctors, an equivalent tune change of $dQ^e_y \approx 0.0008$ was introduced and the beam intensity decay measured for some 20–30 min. After that, the tune correction was turned off for 20–30 min for a "reference" lifetime measurement, which was followed by another ½ hour of TEL-2 operation, and so on. Figure 3.20 shows the total proton intensity lifetime measured for each of these intervals. One can see that initially the beam lifetime improves every time when either TEL2 or the tune correction was introduced. Nevertheless, after some 5 h into the store, the TEL-2 still lead to the lifetime improvement while during two periods of tune correction the lifetime slightly decreased with respect to the unperturbed reference periods (the black bars in Fig. 3.20).

Besides a significant reduction of the proton intensity loss rates, the luminosity lifetime $\tau_L = L/(dL/dt)$ was improved as well. Figure 3.21a compares the changes of

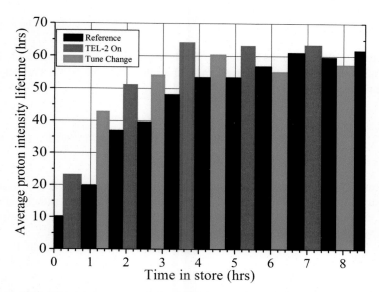

Fig. 3.20 Average proton intensity lifetime $dt/(dN_p/N_p)_{total}$ in store no. 5183 when repetitively, the TEL-2 was turned on protons with 0.3 A of DC electron current (*red bars*), then turned off for reference (*black bars*), next the proton vertical tune was shifted up 0.0008 by quadrupole and sextupole tune correctors (*green bars*), and the correctors were finally turned off again (*black*) [7]

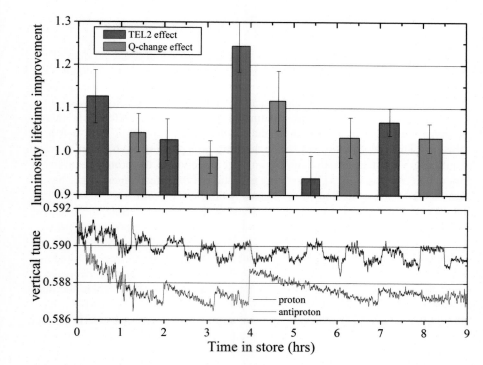

Fig. 3.21 (**a**) *Top*—Luminosity lifetime improvement by TEL-2 and by tune correctors in store no. 5183; (**b**) *Bottom*—the average vertical tunes of the proton and antiproton bunches measured by the 1.7 GHz Schottky detectors [7]

the lifetime of the combined luminosity for the three bunches no. 12, 24 and 36 due to TEL-2 and due to the tune correction in the same store no. 5183. The height of each bar is equal to:

$$R_L = \frac{2\tau_L \left(with\ TEL\ or\ dQ_y\ change\right)}{\tau_L\left(reference\ period\ before\ the\ change\right) + \tau_L\left(reference\ period\ after\ the\ change\right)}.$$

(3.3)

The luminosity lifetime improvement due to TEL-2 is about 12 % at the beginning of the store. Later in the store the TEL-2 effect was somewhat larger than that of the global tune correction dQ_y. The evolution of the average proton and antiproton tunes is shown in Fig. 3.21b.

The TEL induced improvements in the luminosity lifetime of about 10 % are significantly smaller than the corresponding changes in the proton intensity lifetime (about a factor of 2) because the luminosity decay is driven mostly by other factors, the strongest being the proton and antiproton emittance increase due to intra-beam scattering and the antiproton intensity decay due to luminosity burn-off. Usually, these factors combined lead to the decay of instantaneous luminosity approximately given by [1, 2]

$$L(t) = \frac{L_0}{1 + t/\tau_L},$$

(3.4)

so the total integrated luminosity over a store is proportional to the product of the initial luminosity and the luminosity lifetime $L_0\tau_L ln(1 + T/\tau_L)$, where T denotes the store duration. Therefore, a 10 % improvement of the luminosity lifetime τ_L due to TEL-2 results in a proportional increase of the integrated luminosity.

Usually, the proton lifetime, dominated by beam-beam effects, gradually improves with time in a HEP store and reaches some 50–100 h after 6–8 h of collisions. This is due to the decrease in the antiproton population and to an increase in antiproton emittance, both contributing to a reduction of the proton beam-beam parameter ξ^p. In store no. 5119, the time evolution of the effectiveness of the beam-beam compensation was studied by repeatedly turning TEL-2 on a single bunch P12 and off every half hour for 16 h. The relative bunch intensity lifetime improvement R is plotted in Fig. 3.22 from [17]. The first two data points correspond to $J_e = 0.6$ A, but subsequent points were taken with $J_e = 0.3$ A to observe the dependence of the compensation effect on the electron current. The change of the current resulted in a drop of the relative improvement from $R = 2.03$ to $R = 1.4$ A gradual decrease in the relative lifetime improvement is visible until after about 10 h, where the ratio reaches 1.0 (i.e., no gain in the lifetime). At this point, the beam-beam effects have become very small, providing little to compensate. Similar experiments in several other stores with initial luminosities ranging from 1.5 to 2.5×10^{32} cm^{-2} s^{-1} reproduced these results.

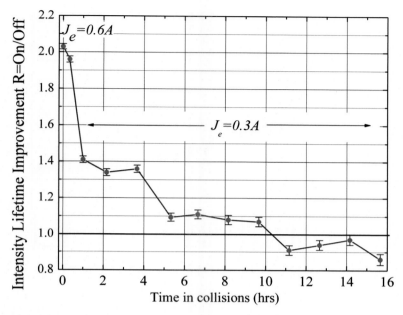

Fig. 3.22 Relative improvement of the TEL induced proton bunch no. 12 lifetime vs. time (Tevatron HEP store no. 119, Dec. 12, 2006, initial luminosity $L = 159 \times 10^{30}$ cm^{-2} s^{-1}) [17]

Comparable improvement of the proton intensity lifetime (up to 40 %) has been observed in experiments performed with TEL-1. The only design difference between the two lenses is that the TEL-1 bending section has a 90° angle between the gun solenoid and the main solenoid while this angle is about 57° for TEL-2. TEL-1 is installed in a location with large horizontal beta-function and mostly shifts horizontal proton tune up. As the proton horizontal tunes are lower for the bunches at the beginning of the bunch trains, P1, P13, and P25 by $\Delta Q_x \approx -(0.002-0.003)$ [1], the TEL-1 effect is the largest for them. The reduction of the global proton loss rates due to the electron lenses can easily be seen by the local halo loss rate detectors installed in the D0 and CDF detectors, which can measure the losses on a bunch-by-bunch basis. Figure 3.23 shows the dependence of D0 proton loss rate on the TEL-1 electron current. In this experiment TEL-1, being a horizontal beam-beam compensation device, was acting on P13 which has the lowest horizontal tune. Bunch P14—unaffected by TEL-1—was chosen as a reference bunch because its behavior in terms of halo and lifetime was very similar to P13, without TEL. The loss rate of P13 dropped by about 35 % once a 0.6 A-peak electron current was turned on, while the P14 loss rate stayed unaffected. After about 12 min the e-current was turned off which made the P13 loss rate return to the reference level. The loss reduction has been repeated several times over the next 4 h in this store and it was confirmed in several other HEP stores.

The dependence of the loss-rate reduction on the electron beam position with respect to the proton beam position has been studied also. Figure 3.24 shows that if

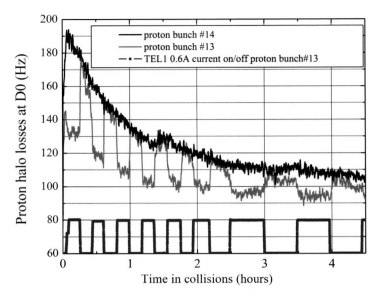

Fig. 3.23 Proton bunch intensity loss rates detected by D0 counters: *black*—for reference bunch 14, *red*—for bunch 13 affected by TEL-1 (first 4 h in store no. 5352 $L = 197 \times 10^{30}$ cm^{-2} s^{-1}) [7]

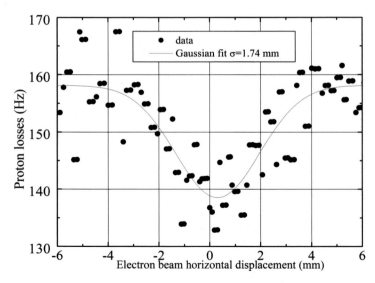

Fig. 3.24 Proton loss rate measured at the D0 detector vs. horizontal displacement of electron beam in TEL-1 [7]

the electron beam is displaced from the proton orbit by more than 4 mm, the effect of the TEL-1 on the D0 proton halo loss rate vanishes.

One can summarize the results of the pioneering long-range beam-beam compensation studies with the Tevatron electron lenses as follows:

(a) Major outcomes of the work are the development of electron lenses, the demonstration of their compatibility with the operation of superconducting hadron collider and the experimental proof of compensation of beam-beam effects in the Tevatron proton-antiproton collider;

(b) The results of the beam-beam compensation studies demonstrate that the shift the proton and antiproton tunes due to electron lenses are as originally predicted in [18] and (3.1). Both electron lenses—TEL-1 and TEL-2—do produce a very strong positive effect on the lifetime of the Tevatron proton bunches which otherwise suffered most from the collisions with antiprotons. The observed lifetime improvement at the beginning of a HEP store (when the beam brightness and luminosity are highest, and the beam-beam interaction is strongest) can be as big as a factor of 2. Only some 10 h into the stores, the beam-beam effects and the BBC gains decrease to insignificant levels. The beam-beam compensation effect is found to be tune dependent and somewhat outperforming the traditional tune correction method. It has to be noted that the difference between two electron lenses—bending angle of the electron trajectory is 90° in one lens and 57° in another—did not significantly impact the reduction of the proton losses by both lenses;

(c) It was experimentally shown that for the successful operation of electron lenses one needs a smooth transverse distribution of the electron current density, a good alignment of the electron beam on the beam of interest— within a fraction of proton or antiproton rms beam size; and low noises and ripples in the electron beam current and position.

(d) We have not seen any sign of coherent instabilities due to the (anti) proton beam interaction with the electron beam, despite initial concerns (see Sect. 3.2.2 above and [18]).

The Tevatron electron lenses were frequently used for the beam-beam compensation in the Tevatron, but were not incorporated in the longer term routine operation of the collider because at some moment, the beam-beam effects had become less severe due to two other improvements—significant reduction of the machine's second order chromaticity $Q'' = d^2Q/d(\Delta p/p)^2$ and intentional controlled blow-up of the antiproton beam size early in the HEP stores to make the antiproton beam wider and better match the proton beam size at the main IPs [2].

3.2 Compensation of Head-On Beam-Beam Effects

3.2.1 Specific Requirements for the Head-On Beam-Beam Compensation

The head-on beam-beam interaction results not only in the shift of the (anti)proton betatron frequencies $\xi = N_p r_p / 4\pi\varepsilon$—see (1.8)—but also in the spread of the frequencies which depend on the amplitude of particles' betatron oscillations J_x, J_y [19]

$$dQ_{x,y}^{P}(J_x, J_y) = \xi \cdot \int_0^1 dt \frac{\left(I_0\left(\frac{J_{x,y}t}{2}\right) - I_1\left(\frac{J_{x,y}t}{2}\right)\right)I_0\left(\frac{J_{y,x}t}{2}\right)}{\exp\left((J_y + J_y)t/2\right)}. \qquad (3.5)$$

where the Gaussian distribution of the opposite bunch is assumed, and the amplitudes (action variables) relate to the normalized particles coordinates $z_n = (x_n, y_n)$ as $z_n = (2J_z\beta_z)^{1/2}\cos(\Phi_z)$, Φ_z is the phase of the betatron oscillations—see Fig. 3.25a. The effect due to Gaussian electron beam will have the same functional dependence on the amplitude but opposite sign because the collisions with electrons focus the protons, while collisions of protons with protons are defocusing. The original proposal for the non-linear beam-beam compensation, therefore called for exact cancellation of the two effects for all particles in the high energy beams and therefore, zero, or close to zero, betatron tune spread $dQ_{x,y}^{P}(J_x, J_y) + dQ_{x,y}^{e}(J_x, J_y) = 0$—see Fig. 2.25b from [20]. From (3.5) and (3.1) one concludes that such

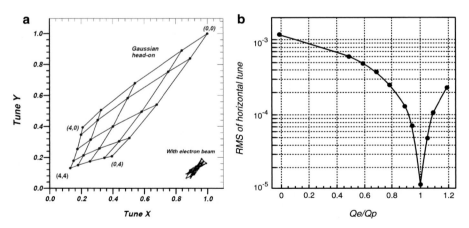

Fig. 3.25 (**a**—*Left*) Antiproton footprint—spread of the vertical and horizontal betatron tunes due to collisions with protons. Tunes are given in units of the head-on beam-beam parameter ξ. Numbers in parentheses show the horizontal and vertical betatron amplitudes in units of the rms antiproton beam size. The smaller footprint in the low right corner is for the case of the non-linear beam-beam compensation by electron lens (displaced for clarity, electron beam distribution slightly aberrant from the Gaussian) [18]; (**b**) RMS of the betatron tune distribution of the protons in the SSC versus the ratio of the electron to proton bunch charges [20]

compensation requires two conditions to be satisfied: (i) the effective number of electrons in the lens is equal to the number of protons $N_e^* = J_e (1 + \beta_e) L_e/(ec\beta_e) = N_p$; and (ii) the electron beam should have the same transverse beam size as the proton beam at the location of the lens $\sigma_e = (\varepsilon\beta_{e\text{-}lens}/\gamma_p)^{1/2} = \sigma_{p,\ e\text{-}lens}$, here $\beta_{e\text{-}lens}$ denotes beta-function at the electron lens. Note that the second condition allows installation of the electron lens essentially anywhere in the ring, and in particular, away from the main IPs which are usually taken by the particle detectors. Moreover, it would be beneficial to install the lenses at the locations with larger beta-functions where σ_e is large as well, because that makes it easier to provide the required electron current density $j_e \sim N_e/\sigma_e$. It was thought that while it is the beam-beam tune spread $\sim\xi$ that makes it impossible to operate colliders with higher intensities within operationally available tune space between major non-linear resonances—e.g., some $\delta Q \approx 0.028$ at the Tevatron working point between tenth and seventh order resonances $Q_{x,y} = 6/10$ and $Q_{x,y} = 4/7$, or $\delta Q \approx 0.033$ in the LHC between $Q_{x,y} = 3/10$ and $Q_{x,y} = 1/3$ absence of the tune—then the beam-beam compensation by appropriate electron lens under the two above conditions (i) and (ii) will allow to increase the maximum possible proton intensities, and therefore, luminosities.

In the following publications [5, 18], it was quickly noted that these two conditions are not sufficient for the most effective compensation. Several other factors need to be taken into account. First of all, the ideal locations for the electron lenses need to have zero or small dispersion D_x desirable to avoid the possibility of synchrobetatron effects $D_x\sigma_{dE/E} < <\sigma_e$, to have equal horizontal and vertical beta functions, and have betatron phase advances from the main IPs to be multiple of 2π: $\Delta\Phi_{x,y} = |\Phi_{e\text{-}lens} - \Phi_{IP}| = k \times 2\pi$, k is integer—as schematically illustrated in Fig. 3.26 from [21]. The beam-beam footprint itself can be significantly distorted by "imperfections" such as a crossing angle at the interaction point or numerous parasitic interactions in multibunch colliders at the locations where two beams do not actually collide but still interact via long-range electromagnetic forces. The

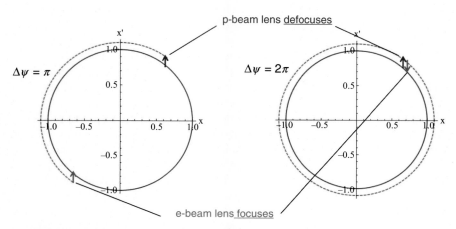

Fig. 3.26 Schematic of head-on beam-beam compensation in a normalized phase space: *left*—with betatron phase advance between electron lens and IP equal to π, *right*—with 2π advance

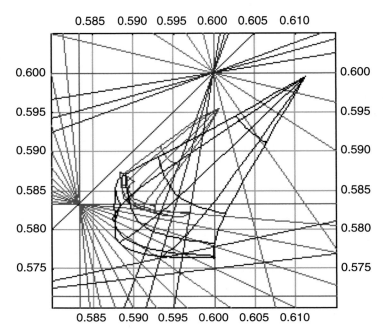

Fig. 3.27 Betatron tune footprint of the Tevatron antiproton bunch footprint without (*larger*) and with beam-beam compensation. Effect of the footprint "folding" is clearly seen at large betatron amplitudes [5]

collider focusing lattice itself is not linear and this should also be taken into account. An additional difficulty is that nonlinearities are numerous and distributed over the collider ring while only one or very few electron beam setups can be installed. For example, results of analytical calculations [5] presented in Fig. 3.27 show how these effects distort the Tevatron antiproton bunch tune footprint without electron lens beam-beam compensation (BBC, larger footprint) and with partial compensation. The bare lattice tunes were set nominally to be $Q_x = 20.585$, $Q_y = 20.575$. The arc lines correspond to equidistant with step 2 values of the total transverse amplitude $A = (A^2_x + A^2_y)^{1/2}$, where $A_{x,y}$ are in the units of antiproton rms size $\sigma_{x,y}$, the radial lines correspond to constant values of A_x/A_y. The straight color lines show sum and difference resonances of orders lower than 13. One can see that the footprint "folding" (when particles with different amplitudes have same betatron tunes) is caused by the long-range interactions with the proton beam. The folding occurs at amplitudes of ~8σ without BBC, and at amplitudes as small as 5σ with BBC. Even very weak high order resonances may lead to a fast particle transport over the region of the folding thus reducing the antiproton beam lifetime. Therefore, this effect could set a natural limit on the maximum degree of the footprint compression. Similarly, the complete footprint compression might pose threat for coherent stability of high energy beams in the presence of impedance. Only partial compensation can be a practical solution to that problem as it allows to leave enough spread of the betatron frequencies to assure the Landau damping of the instabilities.

Another effect deals with different time structures of the beam-beam compensating focusing kick due to the electron lens and that of the defocusing opposite beam of protons. Indeed, in many colliders, the proton bunch length σ_s is comparable to beta function at the interaction point β^* (e.g., about 50 cm and 28 cm, correspondingly, in the Tevatron). Therefore, the betatron phase advance for antiprotons at the main IP is large $\Phi_z = \int ds/\beta(s) \sim \sigma_s/\beta^* \sim 1$. In contrast, the electron beam length of about 2–3 m is much smaller than the beta function at the electron lens compensation setup, and the corresponding betatron phase advance of antiprotons passing the electron beam is very small $\Phi_z \sim 0.01\text{–}0.02$. Thus, the electron beam kick looks like a delta function when transformed to the main IP. Consequently, such a short impact from the electrons contains a lot of resonance harmonics, although the average actions due to proton and electron beams are the same. One can reduce the betatron tune spread with a nonlinear lens, but this alone does not assure the motion is more stable than that with no compensation, because the resonance strengths sometimes can be more important than the tune spread.

The high order nonlinear beam-beam resonance driving terms due to the beam-beam interaction and electron lenses also can be calculated using the Hamiltonian perturbation approach [22, 23]. The Hamiltonian of the collider in action-angle variables (J_z, Φ_z) is given by

$$H\left(J_x, J_y, \Phi_x, \Phi_y\right) = 2\pi J_x Q_x + 2\pi J_y Q_y + \mathrm{Re} \sum_{m,n} h_{m,n}\left(J_x, J_y\right) e^{-in\Phi_x - im\Phi_y}, \quad (3.6)$$

where the order of the resonance is $|m| + |n|$. Modern analytical and numerical modeling tools allow to calculate the resonant driving terms h_{mn} and evaluate their effect on beam dynamics, and particle losses. Figures 3.28, 3.29, 3.30 and 3.31 present the results of such analysis for the RHIC head-on beam-beam compensation project [24]. Major conclusions of those studies are: (a) non-linear beam-beam compensation by electron lenses should allow several fold increase of the luminosity in RHIC polarized proton operation via increase the proton bunch intensity from about $N_p = 1.5 \times 10^{11}$ to 3.0×10^{11}; (b) the electron lenses will effectively compensate the large beam-beam resonance driving terms and large beam-beam tune spread to fit in the current collider tune space between 2/3 and 7/10 resonances if the betatron phase advance between the location of the lenses (IP10) and the main IPs (no. 6 and no. 8) are corrected to be $\Delta\Phi_{x,y} = |\Phi_{e\text{-}lens} - \Phi_{IP}| = (9\pi, 11\pi)$, off the current values of $(8.5\pi, 11.1\pi)$; (c) the frequency map analysis, dynamic aperture, and proton beam loss rate calculations indicate that the optimum compensation for the bunch intensities 2.5 and 3.0×10^{11} is around half beam-beam compensation with a slightly larger electron beam size than the proton beam size (i.e., the electron lens needs to compensate only beam-beam effects of one out of two main IPs, because full head-on beam-beam compensation reduces the proton particle's dynamic aperture and increases the proton beam loss rate); (d) the numerical tracking of the proton beam loss with various beam errors and noises indicated

Fig. 3.28 Calculated tenth order horizontal beam-beam resonance driving terms under different beam-beam conditions in RHIC: in the absence of beam-beam compensation (No BBC), with half-strength BBC, full strength BBC under condition of the phase advances between IP8 and the IP10 electron lens of $\Delta\Phi_{x,y} = |\Phi_{e\text{-}lens} - \Phi_{IP}| = (8.5\pi, 11.1\pi)$, and with corrected phase advance of $(9\pi, 11\pi)$ for half strength BBC. The horizontal axis is the particle's amplitude in units of σ_x. The vertical axis is the resonance driving term in units of $N_p r_0/\gamma_p$ [24]

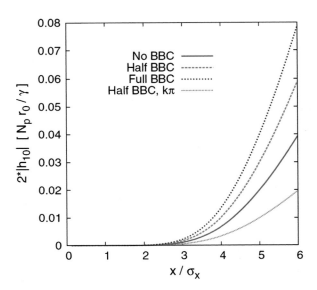

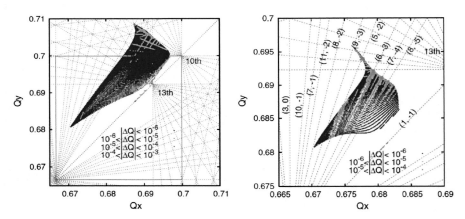

Fig. 3.29 Calculated tune footprints and tune diffusion without (*left*) and with half-strength head-on beam-beam compensation in RHIC. The proton beams collide at IP6 and IP8. The bunch intensity is $N_p = 3.0 \times 10^{11}$ in both cases. Note the scale on the left is about twice the one on the right (0.045 vs. 0.025). Without BBC the tune footprint area goes beyond operational safe conditions (*red box*) and tunes cross strong 7/10 resonances. Situation is greatly improved with half strength BBC [24]

tolerances on the electron beam quality: the electron beam's Gaussian tail cutoff at 2.8σ from the existing electron gun design is acceptable; the random noise in the electron current should be better than 0.1 %; the static and random offsets between the electron and proton beams should not exceed 30 μm and 9 μm, respectively (compare with 310 μm rms beam size—see also Table 2.2).

Fig. 3.30 Calculated dynamic apertures of half head-on beam-beam compensation in RHIC with the betatron phase advance adjusted to $\Delta\Phi_{x,y} = |\Phi_{e\text{-}lens} - \Phi_{IP}| = (9\pi, 11\pi)$ and the global second order chromaticity correction, marked as "Half BBC, $k\pi$, Q" corr. [24]

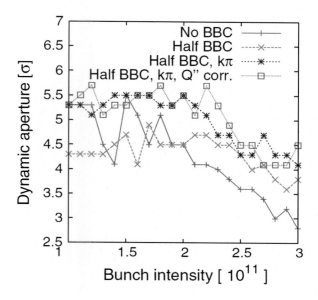

Fig. 3.31 Results of numerical tracking of relative proton beam losses $dN_p(t)/N_p$ in RHIC in 2×10^6 turns for the initial proton bunch intensity $N_p = 3.0 \times 10^{11}$ without beam-beam compensation, with half beam-beam compensation, with the betatron phase advance adjustment, and with the global second order chromaticity correction [24]

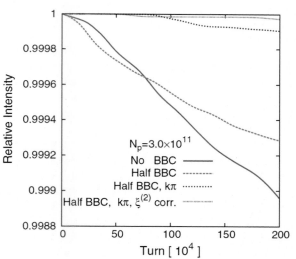

3.2.2 Experimental Studies of the Head-On Beam-Beam Compensation with the Tevatron Electron Lenses

3.2.2.1 Non-linear Beam-Beam Compensation of the Emittance Growth by TELs

Historically, the first experimental indication of non-linear head-on compensation came early in the Tevatron Run II operation when electron lens was employed to suppress emittance growth of antiproton bunches. It is to be noted that though the

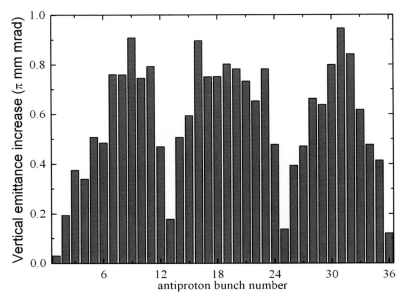

Fig. 3.32 Antiproton bunch emittance increase over the first 10 min after initiating collisions for HEP store no. 3231 with an initial luminosity $L = 48 \times 10^{30}$ cm^{-2} s^{-1} [7]

TEL electron beam profile was Gaussian, the TEL was intended for the long-range compensation and therefore was installed at the location with grossly different vertical and horizontal beta-functions, non-zero dispersion and not at the ideal betatron phase advance from main IPs, i.e., the conditions were far from ideal for the head-on BBC.

Figure 3.32 shows the vertical emittance blowup for all three trains of antiproton bunches early in one of the Tevatron HEP stores in 2003. One can see a remarkable distribution along the bunch train which gives rise to the term "scallops" (three "scallops" in three trains of 12 bunches) for this phenomenon—the end bunches of each train exhibit lower emittance growth than the bunches in the middle of the train. Because of the threefold symmetry of the proton loading, the antiproton emittance growth rates are the same within 5–20 % for corresponding bunches in different trains (in other words, bunches nos. 1, 13, and 25 have similar emittance growths). The effect is dependent on the antiproton tunes; particularly on how close one bunch is to some important resonance. Typically, the Tevatron working points during 2003 operations were set to $Q_x = 0.582$ and $Q_y = 0.590$. At this working point, fifth-order (0.600), seventh-order (0.5714), and twelfth-order (0.583) resonances all play major roles in the antiproton beam dynamics. It was observed that vertical tune changes as small as -0.002 often resulted in a reduction of the amplitude of the "scallops." Smaller but still definite "scallops" were also seen in protons. After the initial 0.5–1 h of each store, the growth rate of each bunch decreased significantly. Such decrease is understood to be due to the steady

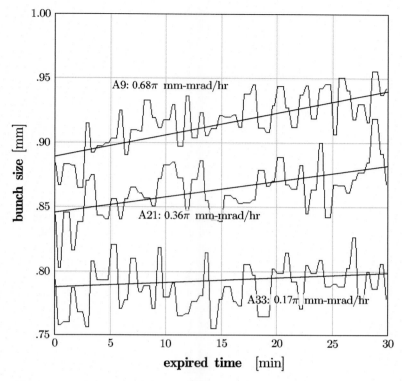

Fig. 3.33 Evolution of antiproton bunch emittances over the first 30 min of HEP store no. 2540: emittances for the ninth bunch in each of three trains are presented: nos. 9, 21, and 33; TEL-1, with the Gaussian electron gun, is acting only on bunch no. 33 [7]

decrease of the antiproton tune shift induced by the protons while the proton beam size grows and the proton intensity rapidly decreases at the beginning of a HEP store.

TEL-1 was used at the beginning of several HEP stores in an attempt to reduce the emittance growth during the first half hour of collisions by acting on a single bunch. The goal was to significantly decrease the emittance growth of the particular bunch with respect to its "sibling" bunches (the equivalent bunches in the other two trains), or with respect to the same bunch in other, similar stores. TEL-1 was timed on a single antiproton bunch at the beginning of the Tevatron stores and the vertical emittance growth of that antiproton bunch was monitored.

Figure 3.33 presents the evolution of the vertical rms sizes of three antiproton bunches (nos. 9, 21, and 33) over the first 34 min after "initiating collisions" in store 2540 (May 13, 2003). The TEL was acting only on bunch no. 33. The bunch size was measured by a Synchrotron Light Monitor [8]. The corresponding emittance growth was $0.68 \pm 0.06\pi$ mm-mrad/h for antiproton bunch no. 9, $0.37 \pm 0.07\pi$ mm-mrad/h for no. 21, but only $0.16 \pm 0.07\pi$ mm-mrad/h for no. 33, as shown by

Table 3.1 Growth of the rms vertical antiproton emittance in the beginning of HEP stores for bunches #9, 21 and 33

Store #	Duration (min)	#9 growth	#21 growth	#33 growth	
#2536	40	1.65	1.53	1.55	TEL-1 off
#2538	35	0.32	0.28	0.46	TEL-1 off
#2540	34	0.68	0.37	0.17	TEL-1 on
#2546	30	0.65	0.32	0.67	TEL-1 on
#2549	26	0.75	0.60	1.18	TEL-1 on
#2551	34	1.12	1.1	1.17	TEL-1 off

All of the growth numbers are in units of π mm-mrad/h, with a typical fit error of ± 0.07 π mm-mrad/h. For the indicated stores, TEL-1 was acting on bunch no. 33

the fitted straight lines in Fig. 3.37. During this experiment, TEL-1 was set to a current of 0.6 A, an energy of 4.5 kV, and an rms beam size of 0.8 mm. Given an interaction length of 2.0 m, the expected maximum horizontal antiproton tune shift is −0.004, and the expected vertical one is −0.001. After 34 min the TEL1 was turned off, and the emittances of all three bunches leveled. Several attempts were made to test this ability, each on a new store during the first short period. These tests, along with pertinent parameters, are summarized in Table 3.1, where each row represents a different store (the designated store number is listed down the left column).

In the three stores listed without TEL-1 in Table 3.1, the emittance growth rate of bunch no. 33 is similar or just slightly larger than that of its siblings. In stores nos. 2546 and 2549, it is still larger. However, in store no. 2540, the growth of bunch no. 33 is significantly less than that of the other two bunches. The differences between consecutive stores are considerable, but the only intentional difference is the application of TEL-1. Soon after store no. 2551, a set of sextupole correction magnets was employed to lower the antiproton tune sufficiently enough (without affecting proton tunes) so the scallops were avoided and there was no operational need in usage of TEL-1 for that purpose anymore.

The effect of TEL-1 in stores no. 2540, no. 2546 and no. 2549 is obvious, though not well-controlled, since it can have an adverse effect instead (store no. 2549, for example). It is believed the reason is insufficiently precise centering of the electron beam on the antiproton orbit: first of all, the antiproton orbit itself changed from store to store by as much as 1 mm at the TEL-1 location; and, secondly, the BPMs used in TEL-1 had an observable 0.5–1.5 mm systematic difference between the nanosecond-scale antiproton bunch and the microsecond-scale electron pulse scales (though the statistical accuracy of either measurement was about 30 μm). Such a large error in the BPM measurement led to difficulties in the experiment repeatability.

In summary, the reduction of the antiproton emittance growth rate in some early BBC studies with TEL-1 was observed, but the effect was not reproduced reliably because of poor control of the electron beam centering on the antiprotons. Later in the Tevatron Collider Run II operation, both the proton and antiproton orbit stability and accuracy of the TEL BPMs were greatly improved (by about a factor of 5–10), but there were no operationally troublesome effects due to the non-linear beam-beam interactions in the Tevatron, and therefore, no more attempts to compensate them.

3.2.2.2   Dedicated Tevatron Studies of Non-linear Beam-Beam Compensation

While an improvement of the Tevatron performance by head-on beam-beam compensation was not foreseen, at the end of the Collider Run II in September 2009 and July 2010, dedicated experimental studies with Gaussian electron lenses [25] had been arranged to prove the feasibility of the concept and to provide the experimental basis for the simulation codes used in the planned application of electron lenses to the RHIC collider at BNL.

It was found that in spite of the very different time structure of the ns-long antiproton bunches and of the electron pulse of about 600 ns, alignment of the electron beam with the circulating beam using a common beam position monitor was accurate to within 0.1 mm and reproducible from store to store. At the nominal collider working point in tune space, the electron lens did not have any adverse effects on the circulating beam, even when intentionally misaligned. To observe any losses due to 5 kV 0.5 A Gaussian electron beam with $\sigma_e = \sigma_a = 0.6$ mm, the antiproton betatron tune was intentionally lowered by -0.003, and dependence of the particles losses on the electron beam position measured and found in excellent agreement with the LIFETRAC numerical code [26, 27] predictions— see Fig. 3.34.

With only antiprotons in the machine, the tune shift and tune spread caused by the electron lens were clearly seen—see Figs. 3.35 and 3.36. Dedicated collider stores with only three proton bunches and three antiproton bunches— i.e., no long-range interactions, head-on effects only—were attempted, but the experimental conditions were not ideal. The data was used for code benchmarking. Moreover, tune scans conducted during these special stores provided a direct comparison between the lifetimes of a control antiproton bunch, a bunch affected by the electron lens, and a bunch experiencing reduced beam-beam forces, as shown in Fig. 3.37. As mentioned above, the Tevatron collider conditions were not ideal for a direct demonstration of the beam-beam compensation concept for two main reasons: head-on nonlinearities for cooled antiprotons were weak during normal operations; and the lattice requirements (zero dispersion, phase advance close to an integer multiple of π) were not exactly met at the electron lens.

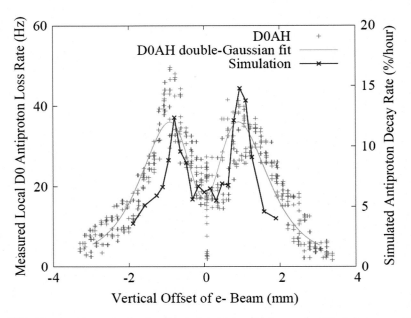

Fig. 3.34 Measured loss rates (*red*) and calculated intensity decay rates (*blue*) during a vertical electron beam scan across the antiproton beam. The antiproton vertical tune was lowered by 0.003 to enhance the effect. No losses caused by the electron beam were observed with nominal antiproton tunes [25]

3.2.3   Significant Improvement of RHIC p-p Luminosity by Gaussian Electron Lenses

3.2.3.1   Initial Experimental Characterizations of the RHIC Gaussian Electron Lens Effects

There are two head-on beam-beam interactions in RHIC at two low-beta $\beta^* = 0.85$ m interaction points IP6 and IP8, as well as four long-range beam-beam interactions IP2, IP4, IP10, IP12 with much larger beta-function $\beta = 10$–20 m and large separation of about 10 mm between the beams. The luminosity in polarized proton operation is severely limited by the head-on effect—as can be seen in Fig. 3.38. At the nominal beam-beam tune-shift parameter $\xi = -0.012$/IP, bunches with two head-on collisions experience a larger proton loss throughout the store than bunches with only one collision. The enhanced loss is particularly strong at the beginning of a store and limits the maximum proton bunch intensity and therefore, luminosity. Partial, half-strength compensation of the head-on beam-beam effect with one electron lens installed in each ring some 3 m from IP10, together with the proton intensity and emittance upgrades aims to approximately double the collider luminosity [21].

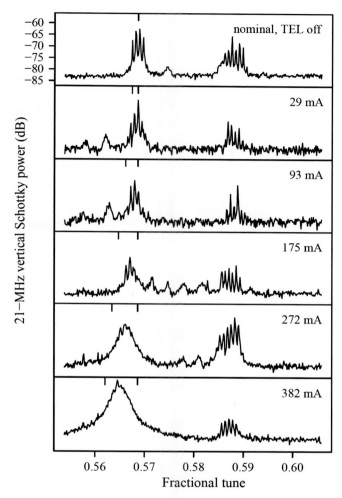

Fig. 3.35 The Tevatron Schottky spectra vs. TEL-2 current. The vertical tick marks indicate the expected magnitude of the linear beam-beam parameter dQ_e, see (3.1). The width of the vertical tune line agrees well with the hypothesis that dQ_e represents the maximum tune shift [25]

In 2014, RHIC operated only with 100 GeV gold beams, cooled by a stochastic cooling system with the emittance cooling time of about 1 h (as is the IBS growth time without cooling) and almost all beam losses were from luminosity burn-off. The Au-Au luminosity was not limited by head-on beam-beam interactions at $\xi_{Au-Au} = -0.006$/IP and compensation was not necessary, but that period was extensively used for the electron lens hardware and electron beam commissioning and initial characterization studies of the electron-hadron beam interaction.

A fast and reliable alignment of the electron beam on the ion beam has been implemented with a novel detector based on measuring backscattered

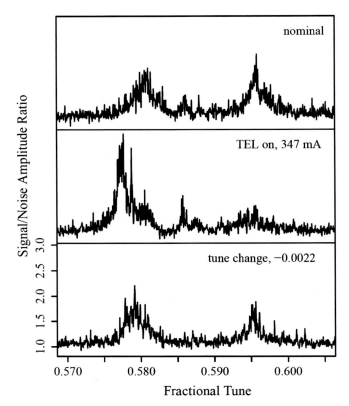

Fig. 3.36 Spectra of transverse coherent modes with and without TEL-2 measured by a high precision bunch-by-bunch system [28]. The top plot shows the spectrum of coherent modes under nominal conditions: the linear beam-beam parameter was $\xi_a = 0.010$ for antiprotons and $\xi_p = 0.0046$ for protons (two IPs). The middle spectrum corresponds to the electron lens acting on the bunch, with $dQ_e = -0.006$. For comparison, the bottom plot shows the effect of lowering the vertical antiproton tune by 0.0022. In the middle plot, one can see a downward shift of the first eigenmode and a suppression of the second [25]

electrons—see its description in Chap. 2.3.3—which gave reliable signals, with counting rates ranging over six orders of magnitude. The signal was used in an application to maximize the overlap region through automatic position and angle scans (Fig. 3.39). With optimized alignment, $J_e = 1$ A $\sigma_e = 0.5$ mm electron beam induced positive Au tune shift of $dQ^e = +0.004$ in both vertical and horizontal planes, close to theoretical model—see Fig. 3.40.

Another important study confirmed that there is no additional emittance growth or beam loss in both pulsed and DC operation of the RHIC electron lens. The stochastic cooling time of about 1 h sets the resolution for the detection of additional emittance growth. Initial emittance growth tests were made with electron beam pulses that covered only two bunches to separate out any effects arising from ion accumulation in the lenses. Conditions could be setup so that no additional

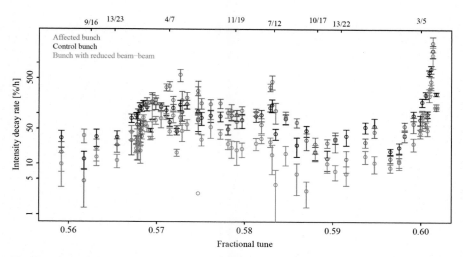

Fig. 3.37 Measured decay rates of the three antiproton bunches during a diagonal tune scan in a special 3-on-3 collider store: the bunch affected by the electron lens with tune shift of $dQ_e = -0.002$ (*magenta*), the unaffected "control" bunch (*dark blue*), and the bunch colliding with the two least dense proton bunches (*green*). Lifetimes and tune space were obviously better for the latter. Some resonances (4/7 and 7/12, for instance) appear stronger with the lens on, whereas the 3/5 is weaker [25]

Fig. 3.38 Intensity evolution of RHIC polarized proton bunches in a 2012 collision run: bunches experiencing collisions in two IPs have significantly higher losses than those with one head-on collision IP [21]

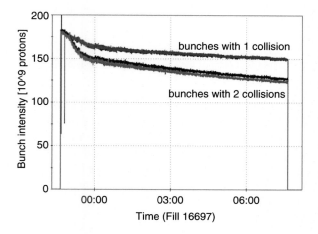

beam loss or emittance growth was observed up to 500 mA electron current. Figure 3.41 shows a test in which the Blue lens electron beam current was successively increased to 800 mA DC, with the electron lens in Yellow off to compare the particle loss rates and emittances. Throughout the current increase, the tunes of the Blue beam were reduced by up to $(-0.004, -0.004)$ to compensate for the additional focusing from the electron lens. The emittance cooled down as fast as

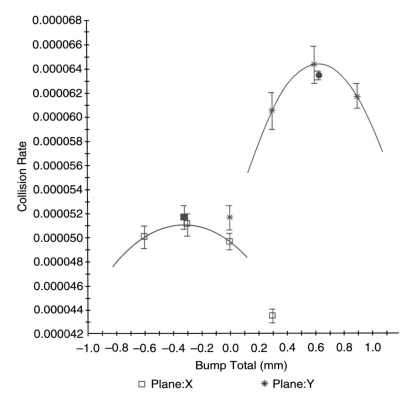

Fig. 3.39 Signal from the RHIC e-lens backscattered electrons monitor used for the automatic transverse alignment of the electron beam on hadron beam [29, 30]

the Yellow emittance, and the Blue loss rate did not exceed the Yellow loss rate by more than 1 %/h after tune correction. For this experiment the Au bunch fill pattern was prepared such that all bunches had two collisions, and the Au and electron beams had been optimally aligned in both position and angle.

Finally, a test was conducted with Au beam, in which the electron lens solenoid main field was lowered from 3 to 1.5 T for which simulations and theory predicted skew-wake instability described in detail above in Chap. 2. No instability was observed in the experiment.

3.2.3.2   RHIC Proton-Proton Luminosity Improvement by Gaussian Electron Lenses

Two Gaussian electron lenses were successfully used for partial compensation of the head-on beam-beam effects during the Relativistic Heavy Ion Collider (RHIC) 100 GeV polarized proton-proton collision run in 2015 [31]. The lenses were

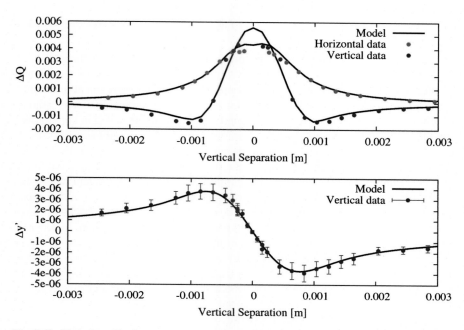

Fig. 3.40 Variation of horizontal and vertical betatron tunes (*top*) and beam orbit (*bottom*) of one of the RHIC 100 GeV gold beams as a function of the vertical position of the electron beam [29]

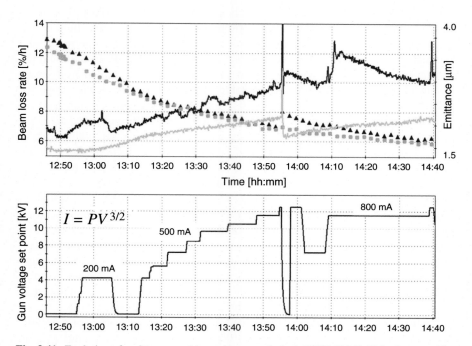

Fig. 3.41 Evolution of emittances and beam loss rates in both RHIC 100 GeV Au beams—*Blue* and *Yellow* (*top*) as a function of the Blue ring electron lens gun anode voltage (*bottom*). Betatron tunes of the *Blue beam* were readjusted by up to $(-0.004, -0.004)$ with increasing electron current [29]

somewhat improved compared to the previous year *Au-Au* run studies [30]: in particular, after improvement of the Blue ring lens's superconducting magnet inner cooling system, both the Blue and Yellow main SC solenoids were ran at the design magnetic field of $B_{main} = 6$ T. A new electron gun with larger diameter cathode (15 mm vs. 8.2 mm previously) was designed, machined and installed in both lenses. The electron beam transverse profiles were measured using a YAG screen systems and fitted with a Gaussian distribution with variable rms size in the lenses $\sigma_e = 0.55$–0.7 mm (typically matched to the rms proton size of about $\sigma_p \approx 0.6$ mm). Significant upgrade of the electron lens control system allowed efficient automatic control of key parameters during operation, for example, optimization of the transverse overlap of the electron and proton beams using the electron backscattering detector in conjunction with an automated proton orbit control program. In order to assure the correct betatron phase advance of integer of π between the electron lenses and the main IP (the PHENIX experiment), a new Achromatic Telescopic Squeeze (ATS) focusing lattice was developed for the proton operation [32]. The lattice has $\beta_{x,y} = 15$ m at the center of each lens where the electron beam and proton beam collide head-on. It also reduces the second order chromaticity, that presumably should help to increase the dynamic aperture of the machine.

The electron lenses were employed in every store over the entire Run 15 *p-p* run, which lasted over 10 weeks. Typical operation sequence was as follows : (a) as soon as the proton beam energy get to 100 GeV, the electron currents in both lenses are brought on while the proton and electron beams are separated horizontally and the proton beams are separated vertically from one another at both colliding IPs; (b) one proton-proton interaction is introduced at the PHENIX experiment and after a short delay of 15 s the other proton collision at the STAR detector and the electron-proton collisions are introduced simultaneously; (c) the electron lenses prevent emittance growth and unacceptable beam losses early in the stores for record high beam-beam parameters $\xi_p \sim 0.011/IP$; (d) 5 kV electron beams are kept on for about 1–1.5 h with the currents stepped down gradually and turned off when the proton beam-beam parameter are decayed to operationally safe values due to natural IBS-driven emittance growth and beam intensity reduction. The electron lenses had typical currents of $J_e \sim 0.6$ A in Blue lens and $J_e \sim 0.8$ A in the Yellow lens, and, therefore, produce positive linear tune shifts in both horizontal and vertical planes $dQ_x/J_e \approx 0.0092/A$ and $dQ_x/J_e \approx 0.0082/A$—see Fig. 3.42 [30]. An automatic machine tune control system was used to maintain the global betatron frequencies $Q_{0(x,y)} + dQ_{x,y}$ at constant values (within $\sim(0.001$–0.002)) to avoid detrimental effects associated with the tune distribution coming too close to dangerous resonances. Detrimental effects due to the electron lenses were found negligible—e.g., it required special, very long ~ 6 h stores with the lenses on to estimate the additional emittance growth induced by the lenses to be about than 0.1 μm mrad/h and lens and additional proton losses of $1 - 2$ %/h [33].

The upgrades introduced in 2015 resulted in smashing improvement of the RHIC performance [34]. While in the previous 100 GeV polarized proton collider run in 2012, strong emittance growth limited the peak collider luminosity at $L_{peak} \sim 50 \times 10^{30}$ cm^{-1} s^{-1}, corresponding to a beam-beam parameter per IP of

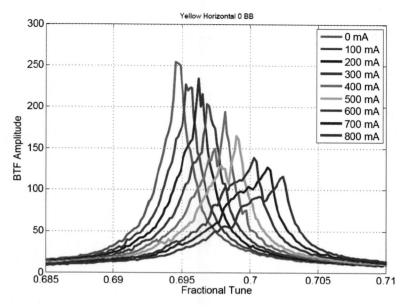

Fig. 3.42 RHIC beam transfer function (BTF) measurements of the horizontal tune distribution vs. current in the *Yellow* electron lens. No proton-proton collision at either STAR or PHENIX experiments, the proton beam collides only with the electron beam. The tune shift of about $dQ_x \approx 0.009/A$ can be clearly seen

$\xi_p = 0.006$; in 2015, with the electron lenses, the beam-beam parameter of 0.011 per IP has been achieved routinely in RHIC stores without the large emittance growth experienced in Run-12. This has led to the 150 % increase in peak luminosity and 90 % increase in average (over store duration) luminosity relative to Run 12—see Fig. 3.43 [31]. Preliminary analysis estimates that the obtained performance improvement could approximately equally attributed to the non-linear beam-beam compensation by the Gaussian electron lenses and to the dynamical aperture gains due to the ATS lattice. Further disentangling of these two effects will be carried out in the future RHIC *p-p* runs. It is also thought that the achieved record luminosities are now limited by the beam brightness set by the RHIC proton injector complex and with corresponding upgrades of the upstream proton accelerators the collider performance with the electron lenses can be further doubled [33].

3.2.4 Potential for the LHC Luminosity Upgrade

The potential of an electron lens to alleviate the head-on beam-beam limit in the LHC was already anticipated as early as 1995 [20]. With the demonstration of feasibility in the Tevatron, it has become of great interest to investigating its benefit for the LHC. In the nominal LHC operating scenario with 1.15×10^{11} protons per

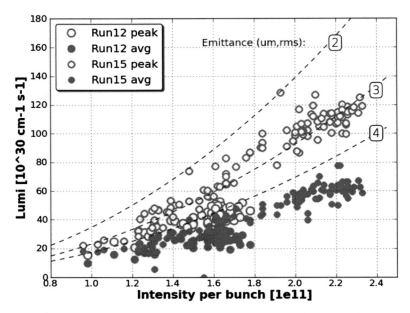

Fig. 3.43 RHIC peak and average *p*-*p* luminosity for all stores in 2012 and in 2015 as a function of proton bunch intensity N_p. Peak luminosity in Run-12 saturated at $L \sim 50 \times 10^{30}\,\mathrm{cm}^{-1}\,\mathrm{s}^{-1}$ due to beam-beam induced emittance blowup. Electron lens compensation and improved off-momentum dynamic aperture allowed several fold higher luminosity in Run-15. The blue dashed lines are curves of constant emittance [31]

bunch, the transverse emittance of 3.75 μm, and 30 long-range collision points per one main IP the total beam-beam tune shift will not exceed $\xi_{tot} = 0.015$ and the beam-beam effects do not cause any particularly adverse effects on the beam dynamics. Situation might get worse with upgraded machine parameters needed for significant luminosity increase. For example, if the bunch intensity is doubles and beam transverse emittance reduced by a factor of 2, the luminosity can reach $10^{35}\,\mathrm{cm}^{-2}\,\mathrm{s}^{-1}$, i.e., ten times the machine design value at 14 TeV c.o.m. In this case, however, the beam-beam effects with the beam-beam paranter as high as 0.03 would become a strong limiting factor and lead to deterioration of the machine performance.

Gaussian electron lens can surely fully or partially reduce the beam-beam tune spread and effectively compress the proton footprint as shown in Fig. 3.44 from [35], but that alone does not guarantee improved beam lifetime.

Numerical particle tracking with weak-strong macro particle code LIFETRAC [27] shows that the efficiency of the electron lens head-on beam-beam compensation in the LHC grows rapidly with the beam-beam tune shift. The numerical model of the simulations [36, 37] uses a three dimensional Gaussian strong bunch and the LHC ring is represented by a series of linear 6D transformations between the IPs. Also included are the first and second order chromaticity, and thin multipoles up to tenth order. The machine lattice is read from MAD-X decks and a full account of

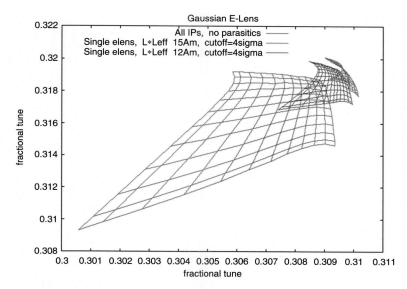

Fig. 3.44 The LHC original (*red*) and compressed footprints for two values of the current in the Gaussian electron lens

coupling and collision sequence is taken. In order to evaluate the beam life time an aperture restriction is placed at the distance of 5–6 σ of the beam and particles reaching this aperture are counted and their coordinates collected. In the simulations reported in [36] the Gaussian electron lens was treated as a thin element placed at the "bbc" section near the LHC IP1, and 10,000 macroparticles particles were tracked over ten million turns in a typical simulation run that corresponds to approximately 900 s of real time. It was found that while the non-luminous fraction of the beam lifetime of the weak beam is very good ~1000 h at the nominal beam-beam parameter value of $\xi = 0.00375$ per IP, it quickly drops to ~100 h for the doubled proton bunch intensity of $N_p = 2.3 \times 10^{11}$ and to ~10 h for the tripled intensity $N_p = 3.45 \times 10^{11}$.

The tracking also demonstrates that it is the combination of long-range and head-on beam-beam effects that causes particle losses, and, e.g., in the absence of long-range collisions the large tune spread generated by the head-on interaction does not lead to significant life time degradation. The effectiveness of the electron lens compensation is found on dependent on the electron current and the rms beam size, number of electron lenses in the ring, and the betatron phase advance between the main IP and the location of the lens—see Fig. 3.45 from [37].

The chart in Fig. 3.46 summarizes the effect of full beam-beam compensation with a Gaussian electron beam on the proton life time for different proton beam intensities. The proton losses are reduced by almost a factor of two for the doubled bunch intensity, and by a factor of four for the tripled intensity. The topic of the effective beam-beam compensation in the LHC is being actively discussed now and several workshops on the subject have taken place [38, 39].

Fig. 3.45 The LHC beam
intensity decay simulations
with the beam-beam
parameter $\xi_{tot} = 0.03$ for
three scenarios of the
Gaussian electron lens
compensation: no lens (zero
current), and two values of
the phase advance between
IP1 and the lens

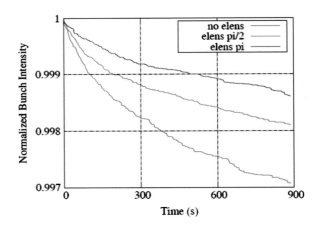

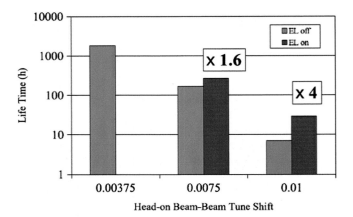

Fig. 3.46 The LHC beam lifetime gain to be expected from the full (100 %) compensation by the electron lens versus beam-beam tune shift—values for the with lens on (*red*) and off (*blue*). The quoted value of the beam-beam parameter is per IP $\xi = \xi_{total}/N_{IP}$

References

1. V. Shiltsev et al., Phys. Rev. ST Accel. Beams **8**, 101001 (2005)
2. V. Lebedev, V. Shiltsev, *Accelerator Physics at the Tevatron Collider* (Springer, New York, 2014)
3. P. Bagley, in *Proceedings of 1996 European Particle Accelerator Conference* (Barcelona, Spain, 1996), p. 1155
4. N.K. Mahale, S. Ohnuma, Part. Accel. **27**, 175–180 (1990)
5. Yu. Alexahin, V. Shiltsev, D. Shatilov, in *Proceedings of 2001 IEEE PAC* (Chicago, IL, USA, 2001), p. 2005
6. D. Shatilov, V. Shiltsev, FERMILAB-TM-2124 (2000)
7. V. Shitsev et al., New J. Phys. **10**, 043042 (2008)

8. R. Moore, A. Jansson, V. Shiltsev, JINST **4**, P12018 (2009)
9. V. Shiltsev, Yu. Alexahin, K. Bishofberger, G. Kuznetsov, N. Solyak, D. Wildman, X.L. Zhang, in *Proceedings of 2001 IEEE PAC* (Chicago, IL, USA, 2011), p. 154
10. A. Kuzmin, A. Semenov, V. Shiltsev, Fermilab beams-doc-842 (2004, unpublished note, https://beamsdoc.fnal.gov)
11. R. Schmidt, in *Proceedings of 3rd Workshop on LEP Performance (CERN, 1993), CERN Preprint* SL/93-19 (1993), p. 139
12. V. Parkhomchuk, V. Reva, V. Shiltsev, Tech. Phys. **48**, 1042 (2003); translated from Zhurnal Tekhnicheskoi Fiziki **73**, 105 (2003)
13. K. Bishofberger, Ph.D. Thesis, University of California, Los Angeles (2005)
14. V. Ranjbar, in *Proceedings of 2005 IEEE PAC* (Knoxville, TN, USA, 2005), p. 1353
15. V. Scarpine, V. Kamerdzhiev, B. Fellenz, M. Olson, G. Kuznetsov, V. Shiltsev, X.L. Zhang, in *Proceedings of 2006 Beam Instrumentation Workshop, Fermilab,* AIP Conf. Proc. 868, ed. by T. Meyer, R. Webber (AIP, Melville, 2006), p. 481
16. X.-L. Zhang et al., Phys. Rev. ST Accel. Beams **11**, 051002 (2008)
17. V. Shiltsev, Y. Alexahin, K. Bishofberger, V. Kamerdzhiev, G. Kuznetsov, X.L. Zhang, Phys. Rev. Lett. **99**, 244801 (2007)
18. V. Shiltsev, V. Danilov, D. Finley, A. Sery, Phys. Rev. ST Accel. Beams **2**(7), 071001 (1999)
19. A. Burov, V. Lebedev, Phys. Rev. ST Accel. Beams **12**, 034201 (2009)
20. E. Tsyganov, A. Taratin, A. Zinchenko, CERN Preprint SL-Note 95-116 (1995)
21. W. Fischer et al., arxiv:1410.5315 (2014)
22. Y. Papaphilippou, F. Zimmermann, Phys. Rev. ST Accel. Beams **5**, 074001 (2002)
23. T. Sen, B. Erdelyi, M. Xiao, V. Boocha, Phys. Rev. ST Accel. Beams **7**, 041001 (2004)
24. Y. Luo et al., Phys. Rev. ST Accel. Beams **15**, 051004 (2012)
25. G. Stancari, A. Valishev, arxiv:1312.5006 (2013)
26. D. Shatilov et al., in *Proceedings of 2005 IEEE PAC* (Knoxville, TN, USA, 2007), p. 4138
27. A. Valishev et al., in *Proceedings of 2005 IEEE PAC* (Knoxville, TN, USA, 2007), p. 4117
28. G. Stancari, A. Valishev, Phys. Rev. ST Accel. Beams **15**, 041002 (2012)
29. W. Fischer et al., in *Proceedings of IPAC 2014* (Dresden, Germany, 2014), p. 913
30. X. Gu et al., in *Proceedings of IPAC'2015* (Richmond, VA, USA, 2015), p.3830
31. V. Schoefer et al., in *Proceedings of IPAC'2015* (Richmond, VA, USA, 2015), p.2384
32. S. White, W. Fischer, Y. Luo, BNL Preprint C-A/AP/519 (2014)
33. W. Fischer, X. Gu, presented at the *Joint HiLumi-LARP Meeting and 24th LARP Collaboration Meeting* (Fermilab, 11–13 May 2015); https://indico.fnal.gov/conferenceDisplay.py?confId=9342
34. See, e.g., at http://www.bnl.gov/newsroom/news.php?a=11715
35. A. Kabel, in *Proceedings of US LARP Collaboration Meeting-12* (08–10 Apr. 2009, Napa, CA); http://larpdocs.fnal.gov/LARP-public/DocDB/DisplayMeeting?conferenceid=67
36. A. Valishev, V. Shiltsev, in *Proceedings of 2009 IEEE PAC* (Vancouver, Canada, 2009), p. 2567
37. A. Valishev, in *Proceedings of IPAC'2010* (Kyoto, Japan, 2010), p. 2071
38. *US LARP Mini-Workshop on Beam-Beam Compensation* (SLAC, 2–4 July 2007), Workshop website: http://www-conf.slac.Stanford.edu/larp/; see summary in W. Fischer et al., in *Proceedings of the CARE-HHH-APD Workshop BEAM07* (Geneva, Switzerland), CERN-2008-005 (2008), p. 12
39. *US LARP Mini-Workshop on Electron Lens Simulations* (BNL, December 24, 2008); see summary in A. Valishev, Y. Luo, W. Fischer, BNL Preprint C-A/AP/353 (2009)

Chapter 4
Electron Lenses for Halo Collimation

As discussed in Chap. 1, particle collimation has become one of the most challenging issues for high energy proton superconducting colliders. The SSC, Tevatron, RHIC, LHC and all envisioned future supercolliders put the issue in the core of their design. The demands of high luminosity result in the increasing beam power in a smaller cross section area of higher energy beams. Even a tiny loss of the circulating beam could lead to catastrophic consequences due to the damage to accelerator and detectors if not carefully controlled. The record high beam power density (power per unit beam area) makes it an enormous challenge to collimate such beams. Following pioneering demonstrations of the transverse and longitudinal hadron beam halo collimation by electron beams, electron lenses are now considered as an additional, very flexible and effective element of any future high energy hadron collider.

4.1 Transverse Collimation: Hollow Electron Beam

The concept of the hollow electron beam collimation [1, 2] has naturally come out after initial studies with the Tevatron electron lenses which showed that sharp edges in the transverse electron beam current distribution lead to fast losses of the (anti) protons traversing them, while the particles passing inside the electron beam had a much longer lifetime—see discussion in Sect. 3.1.2. As depicted in Fig. 4.1, an ideal round hollow electron beam has no electric or magnetic fields inside and strong nonlinear fields outside. If such electron beam encloses the circulating beam of (anti)protons, it would kick halo particles transversely and leave the beam core unperturbed. The speed of diffusion of the halo particles can be greatly enhanced if the electron current varies in time, randomly or in sync with the betatron oscillations. The concept has been successfully demonstrated in the Tevatron collider [3], and its remarkable efficiency led to initiation of the development program to establish the Hollow Electron Beam Collimation (HEBC) for scraping high-power beams in the Large Hadron Collider [4].

© Springer Science+Business Media New York 2016
V.D. Shiltsev, *Electron Lenses for Super-Colliders*, Particle Acceleration
and Detection, DOI 10.1007/978-1-4939-3317-4_4

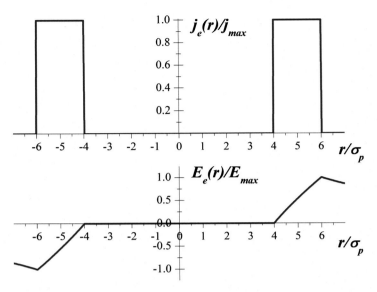

Fig. 4.1 (*Top*) Electron current density distribution in the hollow electron beam; (*bottom*) electric field of the hollow electron beam collimator

4.1.1   Hollow Electron Beam as a Collimator

Figure 4.1 shows the geometry of the radially symmetric hollow beam. The transverse kick $\theta(r)$ experienced by ultra-relativistic particles of magnetic rigidity $(B\rho)_p = \beta_p \gamma_p m_p c^2/e$, $\beta_p \approx 1$ traversing a hollow electron beam at a distance r from its axis depends on the enclosed electron current J_e and on the length L_e of the interaction region:

$$\theta(r) = \theta_{max} \begin{cases} 0 & \text{if } r < r_{min} \\ \dfrac{(r^2 - r^2_{min})r_{max}}{r(r^2_{max} - r^2_{min})}, & r_{min} < r < r_{max} \\ \dfrac{r_{max}}{r}, & \text{if } r > r_{max} \end{cases} \qquad \theta_{max} = \frac{2L_e J_e}{(B\rho)_p r_{max} c} \cdot \left(\frac{1 \pm \beta_e}{\beta_e}\right),$$

$$(4.1)$$

where the sign $+$ or $-$ in the last factor is for situations where electric and magnetic forces are additive or subtractive to each other. Typical angular kick value achievable in the supercolliders is of the order of 0.3 μrad—see, e.g., all relevant parameters for Tevatron and LHC in Table 4.1; that is much smaller than the rms scattering angle in the colliders' primary collimators $\theta_{coll} \approx 17$ μrad in the Tevatron (5 mm thick tungsten) or 3.4 μrad in the LHC's 0.6 m long carbon jaw. What makes the HEBC effective is that it does operate "smoothly", over many turns—every

Table 4.1 Parameters of the hollow electron beam collimation (HEBC) in the Tevatron proton-antiproton colliders and in the LHC—proton beam magnetic rigidity $(B\rho)_p$, electron beam current J_e, energy U_e, maximum radius r_{max}, length L_e and maximum HEBC angular kick θ_{max} in comparison with the rms spread due to passage through the machine's primary collimator θ_{coll}

	$(B\rho)_p$, Tm	J_e, A	U_e, kV	r_{max}, mm	L_e, m	θ_{max}, μrad	θ_{coll}, μrad
Tevatron	33,267	1	5	3	2.0	0.3	17
LHC	23,333	5	10	2.5	3.0	0.3	3.4

time when a particle appears beyond the boundary of the electron beam, it gets a small radial kick.

The electron beam system for the HEBC would be similar to that of the electron lenses, i.e., employ hollow thermionic cathode electron gun (see Fig. 4.2a), a system of solenoid magnets for the beam size compression, electron collector, etc. Uniform and axially symmetric electron current density distribution from the cathode can be optimally preserved in a linear configuration, as shown in Fig. 4.2b from [5].

The diffusion model [6] explains how the selective nature of the electron beam action—only on the particles with large enough betatron amplitudes—could benefit the task of particle collimation in supercolliders. The (anti)proton beam distribution function $f(J_z,t)$ depends on the particle amplitude—action J_z (see (3.5)), where z stands for x or y, it varies in time and usually has a core with long tails (the halo)—as illustrated in Fig. 4.3 from [7]. In the simplest one dimensional case the dynamics of the distribution function follows the equation:

$$\frac{\partial f(J_z,t)}{\partial t} = \frac{\partial}{\partial J_z}\left(D(J_z)\frac{\partial}{\partial J_z}f(J_z,t)\right), \tag{4.2}$$

The diffusion coefficient $D(J_z) = <\Delta J_z^2>/\Delta t$ is an increasing function of action amplitude and is usually due to the nonlinearities of the machine focusing lattice and of the beam-beam interactions. The collimator position define the aperture of the machine at $J_{z,max} = J_c$, or the point where the population density is practically zero. In the diffusion model, the particle loss rate is the flux at the collimator which is proportional to the product of diffusion rate and density gradient:

$$\frac{\dot{N}_p}{N_0} = \frac{d\int f(J_z,t)dJ_z}{dt} = D(J_c)\frac{\partial f(J_z,t)}{\partial J_z}\bigg|_{J_c}. \tag{4.3}$$

With the hollow electron lens, one aims at enhancing diffusion of the tails, reducing their population, as the distribution near the collimator scales inversely with the diffusion coefficient D:

$$f(J_z)\big|_{J_z<J_c} \propto \frac{J_c - J_z}{D(J_c)}. \tag{4.4}$$

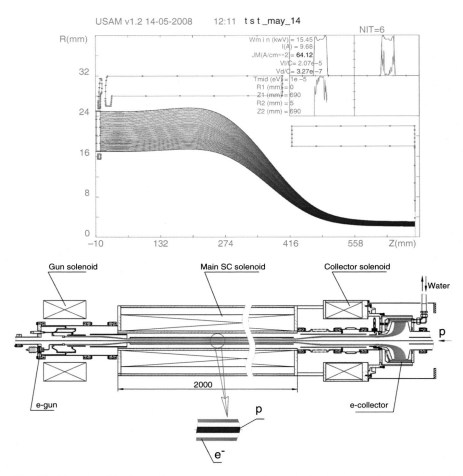

Fig. 4.2 (**a**)—*Top*: Simulated electron trajectories and current density profile (inset) of axially symmetric hollow cathode electron gun; (**b**)—*bottom*: one of possible configurations of the device for the hollow electron beam collimation (HEBC)

Given the flexibility in the electron current control, one can envision three types of the diffusion enhancement: (i) with DC electron current; (ii) with the current, and therefore, the kick, randomly changed from turn to turn; and (iii) with the electron current waveform in resonance with the particle's betatron oscillations. In the first case, the stationary fields of the electron beam which have strong non-linearities—see, e.g., (4.1)—drive the nonlinear resonances the same way as beam-beam interaction does, so the halo particles slowly diffuse to large amplitudes and die on the collimator aperture. Much faster rates of the removal can be achieved with randomly fluctuating kicks $\theta_{max}(t)$. In such case, diffusion can be estimated as:

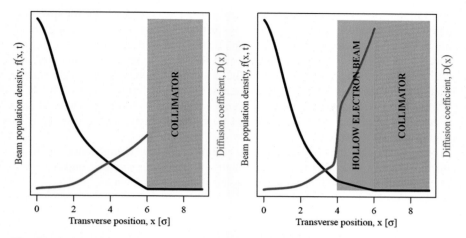

Fig. 4.3 Schematic representation of the diffusion model of collimation with and without the hollow electron beam (from [7]). In this particular case, the primary collimators are shown set at 6σ from the beam orbit, where $\sigma = (2J_z\beta_z)^{1/2}$ is the rms proton beam size at the collimator location, while hollow electron beam has $r_{min} = 4\sigma$ and $r_{max} = 6\sigma$

$$D(J_z) = \frac{dJ_z^2}{dt} \approx f_0 \left(\frac{\beta_z < \theta_{max}^2 >}{2} \right)^2 \sqrt{\frac{J_z - J_{min}}{J_z}}. \qquad (4.5)$$

where the last factor reflects the fact the HEBC acts only when the particle crosses the electron beam wall which happens very rarely at the betatron amplitudes $z = (x,y)$ just above the minimal one $z^2/2\beta_z > J_{min} = r_{min}^2/2\beta_z$. As we will see in the following sections, even in the DC mode, with reasonably high electron currents the HEBC device can enhance the halo-only diffusion in the machines like the Tevatron or LHC by about an order of magnitude, while random kicks could provide further one to two orders of magnitude enhancement. This seems to suffice for most realistic operational requirements, but if much faster halo cleaning is needed, the resonant excitation can be employed.

What makes the HEBC so attractive, and, some believe -indispensable, for the future supercolliders, including the LHC upgrades and the FCC, is its unique combination of very much desired operational advantages: (a) the reduced halo population will decrease the risk of the collimator damage in case of catastrophic events like sudden beam blowups or very fast orbit jumps or fast failures of the components like crab-cavities, and effectively eliminate the loss spikes during the collimator setup or due to the beam jitter; (b) such a collimator is unique as it works for both ion and proton beams due to its purely electromagnetic nature, i.e., there are no nuclear interactions and no ion breakup; (c) the transverse kicks are small and tunable, so that the device acts more like a "soft collimator" or a "diffusion enhancer," rather than a hard aperture limitation; the electron beam could even become an indestructible primary collimator with either higher electron currents or with random or resonant excitation schemes; (d) the HEBC allows for machine impedance reduction either by replacing primary collimators or placing them farther away from the beams; (e) a magnetically confined electron beam can be placed very close to, and even overlap with the circulating beam; moreover, no mechanically moving parts are involved in the collimator alignment as the electron beam positioning needs only electro-magnetic correctors; (f) the electron beam is "refreshable", no high energy beam incident can damage it, contrary to when metal or carbon collimators are employed; (g) thus, no expensive collimator damage diagnostics is needed.

Finally, though the HEBC devices require complex components, such as superconducting solenoids, magnet and high-voltage power supplies, and cryogenics, they rely on the established technologies of the electron lenses.

4.1.2   Experimental Studies of the Hollow Electron Beam Collimation in the Tevatron Collider

The Tevatron electron lenses, though not a perfect fit for the HEBC due to their U-shape, rather than straight geometry, offered an opportunity for initial experimental studies of the hollow electron beam collimation concept. Such studies were performed in the Tevatron between October 2010 and September 2011 and concluded in successful demonstration of this new halo control method. Below we describe these pioneering results, mostly following [3, 7].

In the TEL, the electron beam is generated by a pulsed 5-kV electron gun and transported with strong axial magnetic fields. Its size in the interaction region is controlled by varying the ratio between the magnetic fields in the main solenoid and in the gun solenoid. Halo particles experience nonlinear transverse kicks and are driven towards the collimators. As noted above, under the conditions of axially symmetric current distribution there are no electric or magnetic fields inside the hollow electron beam and the beam core is unperturbed, while particles in the halo of the 980-GeV antiproton bunches get a radial kick up to 0.3 µrad (with a peak electron current of 1 A, an overlap length of 2 m, and a hole radius of 3 mm). A 15-mm-diameter hollow electron gun was designed and built (Fig. 4.4c–e). It was based on a tungsten dispenser cathode with a 9-mm-diameter hole bored through the axis of its convex surface. The peak current delivered by this gun is 1.1 A at 5 kV. The current density profile was measured on a test stand by recording the current through a pinhole in the collector while changing the position of the beam in small steps. The gun was installed in one of the Tevatron electron lenses (TEL-2) in August 2010—see Fig. 4.4b. The pulsed electron beam could be synchronized with practically any bunch or group of bunches. The experiments were carried out with the electron beam acting on selected antiproton bunches with the same electron current on every turn ("continuous mode" of operation, similar to the DC mode discussed above but only for one or a few bunches). Antiprotons were chosen for two main reasons: their

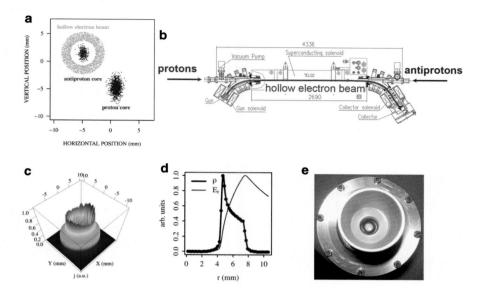

Fig. 4.4 The Tevatron hollow electron beam collimator: (**a**) transverse beam layout in TEL-2; (**b**) top view of the beams in the Tevatron; (**c**) measured electron current profile; (**d**) measured charge density ρ and calculated radial electric field E_r; (**e**) photograph of the electron gun with a 15-mm hollow cathode in the center (from [3])

smaller transverse emittances (achieved by stochastic and electron cooling in the Antiproton Source accelerators) made it possible to probe a wider range of confining fields and hole sizes; and the betatron phase advance between the electron lens and the absorbers was also more favorable for the antiproton collimation. Fig. 4.4a shows in scale transverse cross sections and positions of three beams in the TEL-2. Losses generated by the electron lens were mostly deposited in the collimators, with little effect on the CDF and D0 detector backgrounds. Alignment of the beams was done manually, with a setup time of about 15 min. Alignment is crucial for HEBC operation, and the procedures based on the electron-lens beam-position monitors were found to be reliable in spite of the different time structure of the electron and (anti)proton pulses. No instabilities or emittance growth were observed over the course of several hours at nominal antiproton intensities (10^{11} particles/ bunch) and electron beam currents up to 1 A in confining fields above 1 T in the main solenoid. Most of the studies were done parasitically during regular collider stores.

The particle removal rate was measured by comparing bunches affected by the electron lens with other control bunches. In the experiment presented in Fig. 4.5, the electron lens was aligned and synchronized with the second antiproton bunch train, i.e. on bunches A13-A24, and then turned on and off several times over 3 h at the end of a collider store. The electron beam current was about 0.4 A and the radius of the hole was varied between $6\sigma_y$ and $3.5\sigma_y$, $\sigma_y = 0.57$ mm being the vertical rms antiproton beam size. The black trace is the electron-lens current. One can clearly see the smooth scraping effect. The corresponding removal rates

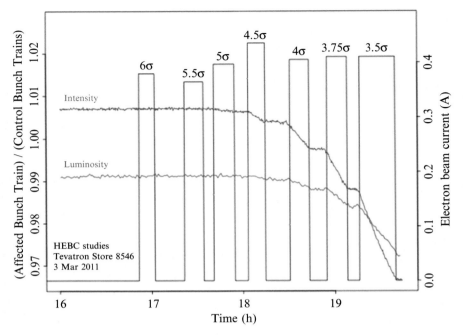

Fig. 4.5 Scraping effect of the hollow electron beam acting on 980 GeV antiproton beam in the Tevatron at the end of a regular collider store for different values of the hole radius, from $6\sigma_y$ to $3.5\sigma_y$. Shown are intensity (*magenta*) and luminosity (*cyan*) of the affected bunch train normalized to the intensity and luminosity of the control (unaffected) train of antiprotons, respectively. The electron beam current (*black trace*) was turned on and off several times

are about a few percent per hour and scale up with reduction of the electron beam hole size.

There is a concern related to adverse effects on the core of the circulating beam. This is a concern because the overlap region is not a perfect hollow cylinder due to asymmetries in gun emission, to evolution under space charge of the hollow profile, and to the bends in the transport system (where electrons are present when antiprotons pass through TEL-2, so antiprotons penetrate the electron beam wall two times—at the entrance and at the exit of the electron lens). The problem was approached from several points of view. First, one can see from Fig. 4.5 that no decrease in intensity was observed with large hole sizes, when the hollow beam was shadowed by the primary collimators. This implies that the circulating beam was not significantly affected by the hollow electron beam surrounding it, and that the effect residual fields near the axis and in the bending sections of the electron lens on the antiproton beam intensity was negligible. Secondly, no difference in emittance growth was observed for the affected bunches: if there was emittance growth produced by the electron beam, it was much smaller than that driven by the other two main factors, namely intrabeam scattering and beam-beam interactions. Finally, negligible effect of the HEBC on the antiprotons in the beam core was confirmed in the collimation position scan: as soon as the primary beam collimator moved closer to antiprotons than the radius of the electron beam hole (i.e., the collimator shadowed the electron beam), the relative intensity decay rate of the affected bunch train went back to the value it had when the lens was off, and became the same as for other, unaffected bunch trains. Even with the HEBC hole size of $3.5\sigma_y$, the effects of the residual fields on the core appeared to be negligible.

The effect of halo removal can also be observed by comparing beam scraping with the corresponding decrease in luminosity. Luminosity is proportional to the product of antiproton and proton populations, and inversely proportional to the beam emittance $L \sim N_a N_p / \varepsilon$, therefore, one would expect for the relative changes to be related as:

$$\frac{\Delta L}{L} = \frac{\Delta N_a}{N_a} + \frac{\Delta N_p}{N_p} - \frac{\Delta \varepsilon}{\varepsilon}. \qquad (4.6)$$

If antiprotons are removed uniformly and the other factors are unchanged, luminosity should decrease by the same relative amount $\Delta L/L = \Delta N_a/N_a$. If the hollow beam causes emittance growth or proton loss, luminosity should decrease even more. What was observed in reality is the opposite—the relative change in luminosity was much smaller than the change in the intensity, that is a clear indication of the scraping of halo, i.e., the particles not contributing to the luminosity—see Fig. 4.6. Also, the ratio between luminosity decay rates and intensity decay rates increased with decreasing the electron beam hole size.

Fig. 4.6 Relative decay rates with hollow electron beam with different values of the hole radius, from 6 to $3.5\sigma_y$: for antiproton intensity $(dN_a/dt)/N_a$ (*magenta*) and for the Tevatron collider luminosity $(dL/dt)/L$ (*cyan*)

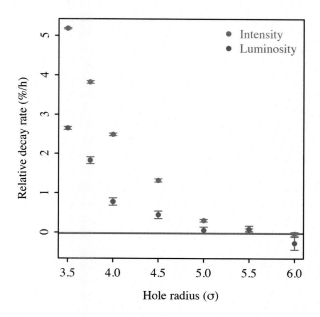

Fig. 4.7 Simultaneous
measurement of local
antiproton losses (gated
beam loss monitor signals)
in response to an outward
50 µm step of the F48
secondary antiproton
collimator: *blue line*—for
the control bunch train,
magenta line—for the
bunch train collimated by
the hollow electron beam
(see text, from [7])

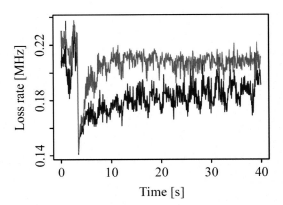

A practical technique to measure the particle diffusion rate $D(J_z)$ in high energy proton beams was suggested in [6]. It is based on analysis of the transients of the beam loss monitor (BLM) signals after small collimator steps. The intercepting (primary) particle collimator motion results either in fast increase of the BLM signal if it moves into the beam, or in a decrease of the signal if the collimator moves out. In both cases, after the initial spike (up or down) the loss rate approaches some equilibrium after the time that scales inversely with the diffusion rate (faster the diffusion—shorter the time). Figure 4.7 depicts time variation of the antiproton beam losses after a fast 50-µm collimator step outwards the beam in the Tevatron. Mathematical processing of such waveforms allows for the determination of the particle diffusion rate $D(J_z)$ using recipes described in [6, 8].

The measurement of the antiproton diffusion coefficient $D(J_y)$ with HEBC as a function of amplitude J_y had been carried out at the end of a regular Tevatron collider store [8]—see Fig. 4.8. The diffusion coefficient is plotted as a function of the vertical collimator position (expressed in terms of the r.m.s. beam size σ_y), for different values of the electron beam current. One can see a clear diffusion enhancement (up to two orders of magnitude for a beam current of 0.9 A) in the region of transverse space where the electron beam is present (electron beam wall begins at some 4σ and extends to about 6.7σ).

Due to greatly enhanced diffusion in the halo, the HEBC reduces the halo population and thus, greatly mitigated the effects of the beam orbit jitter. In the Tevatron, beam orbits oscillate coherently over a wide range of frequencies with the largest amplitudes of up to a few tens of microns at low frequencies (from sub-hertz to a few hertz), due to mechanical vibrations and ground motion [9]. That causes periodic bursts of losses at the beam aperture restrictions (collimators), with peaks exceeding a few times the average loss rate. When the beams are too close to the collimators, such loss spikes are larger and can cause quenches in the superconducting magnets or damage electronic components of the HEP detectors. In March 2011, scintillator paddles were installed downstream of one of the antiproton secondary collimators (F48) to measure the loss spikes and the effects of the hollow electron beam. These loss monitors could be gated to individual

Fig. 4.8 Effect of the
hollow electron beam on the
transverse diffusion
coefficient $D(J)$, as a
function of vertical
collimator position. *Grey
lines* represent the
calculated geometrical
projection of the hollow
electron beam, from 4σ to
about 6.7σ (from [8])

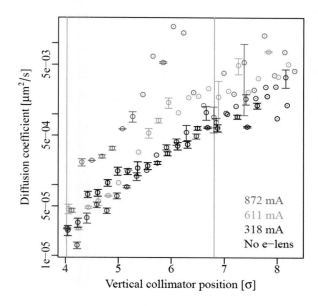

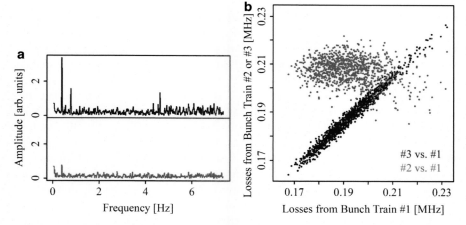

Fig. 4.9 (a)—*left*: Frequency spectrum of the steady-state antiproton loss signals: from control
bunch train No. 1 (*blue*) and from the bunch train No. 2 affected by the hollow electron beam
(*magenta*); coherent peaks due to vibrations are suppressed. (*right*) Correlation between loss rates:
No. 3 vs. No. 1 (*blue*), showing that beam jitter dominates loss fluctuations; No. 2 vs.
No. 1 (*magenta*) shows that the correlation is removed and the peaks due to the beam orbit jitter
are mitigated (from [7])

bunch trains. Figure 4.7 shows the periodic losses due to beam jitter in the control
(unaffected by electron lens) antiproton bunch train. The loss rate variation is much
smaller in the affected bunch train—an indication of halo removal. This effect is
also apparent in the Fourier spectrum of gated losses (Fig. 4.9a). The peaks in the

spectra correspond to notable mechanical vibrations induced by the Main Injector acceleration ramp (0.3 Hz) or by the compressors of the Central Helium Liquefier (4.6 Hz). The electron beam acting on the second bunch train reduced the tails of the antiproton transverse distribution and suppressed these periodic losses. Correlation analysis of their BLM signals further illustrates that—see Fig. 4.9b. The blue points in Fig. 4.9b (right) represent the losses coming from train No. 3 (bunches A25-A36) vs. those from train No. 1 (bunches A1-A13)—both are the trains not affected by TEL-2—which are highly correlated (spread over straight line). Most of the loss variations are not random, but coherent, and can be attributed to the beam orbit jitter. The hollow electron beam eliminates this correlation: average losses are slightly increased, instantaneous losses are randomized, and the spikes are reduced. This phenomena can be interpreted as enhanced diffusion leading to decrease in the antiproton tail population, which in turn translates into greatly reduced sensitivity to beam jitter.

4.1.3   Design Studies of the Hollow Electron Beam Collimation for LHC

Experiments at the Tevatron have convincingly demonstrated that the hollow electron beam collimation is a very effective method for the halo scraping in high energy and high power beams in storage rings and colliders. Naturally, the extension of the HEBC technique to the Large Hadron Collider has been actively studied recently and showed a lot of promise [4, 5, 10–12].

The LHC collimation system [13, 14] performed very well during the 2009–2012 collider operation at about half of the design beam energy (3.5–4.0 TeV vs. 7.0 TeV), providing the machine with an efficient cleaning system [15, 16]. Nonetheless, when trying to push the LHC to and over its design energy and/or luminosity limits, the collimation system needs to boost its performance. It was shown [4] that due to several effects, the present collimation system or its standard upgrade cannot easily assure active and smooth halo scraping during high intensity operation under the LHC luminosity upgrade parameters [17] but that can be addressed by the HEBC system. The effects are: (a) higher sensitivity to spikes due to beam jitter or movement; (b) coherent instabilities due higher impedance of the numerous collimator jaws which need to be placed closer to the beam, to protect the IR focusing magnets allowing reduction of the beta-functions at the high-luminosity IPs; (c) growing probability of the damage to the collimators due to incidents with higher energy, higher intensity and smaller emittance beams; (d) the need for additional protection against possible failures of some existing or novel systems, such as the crab-cavities.

In the HEBC system, the LHC primary collimators will be placed at around $6\sigma_p$ from the beam axis. To effectively scrape the halo of a 7-TeV proton beam, the inner radius of the electron beam in the interaction region r_{min} needs to be placed at about $4\sigma_p$ of the LHC proton rms beam size $\sigma_p \approx 0.32$ mm at the candidate locations

($\beta_{x,y} = 200$ m, the nominal normalized rms emittance $\varepsilon_p = 3.75$ µm). In the case of elliptical proton beams, the HEBC scraping is possible with orbit bumps or by displacing the electron beam. High magnetic fields in the guiding solenoids are needed for the stability and the electron beam transport efficiency. Based upon the TEL experience and technical feasibility, the fields in the gun, main (superconducting), and collector solenoids are set in the ranges 0.2 T $< B_{gun} < 0.4$ T, 2 T $< B_{main} < 6$ T, and 0.2 T $< B_{coll} < 0.4$ T, respectively. That implies magnetic compression factor $k = (B_{main}/B_{gun})^{1/2}$ in the range $2.2 < k < 5.5$, which sets the required sizes of the cathode inner and outer radii. The 1-in. cathode electron gun has been built for this purpose. It has an inner radius of $R_{gi} = 6.75$ mm and outer radius of $R_{go} = 12.7$ mm and demonstrated the expected perveance of 5.3 µP and maximum current over 5 A at 10 kV [18, 19]. According to the magnetic compression relation $R^2_{gi}B_{gun} = r^2_{min}B_{main}$, therefore, these radii translate to 1.2 mm $= 3.9\sigma_p < r_{min} < 9.5\sigma_p = 3.0$ mm and 2.3 mm $= 7.3\sigma_p < r_{min} < 18\sigma_p = 5.7$ mm in the interaction region inside the main solenoid.

There are three options for the geometry of the HEBC device—U-shape geometry (like in the case of the TELs where both gun and collector solenoids are off the proton orbit and on the same side), S-shape (same, but gun and collector solenoids are on the opposite sides of the proton beam—see Fig. 4.10 from [20]) and straight geometry as in Fig.4.2b. The first two geometry options are simpler from the technology perspective and their operations have been demonstrated in the Tevatron and RHIC, but have disadvantages because of the bend areas where proton beam would interact with separated electron beam. The option of the straight HEBC system is free of that deficiency but requires new engineering designs of the electron gun and collector which should allow significant openings so the proton beams can pass through them. The electron current in the HEBC electron gun is easily controlled by the anode-cathode voltage. This flexibility opens up the possibility to operate the hollow electron beam in different modes: continuous— with either DC current or with the same current pulse is delivered on every turn for a subset of proton bunches of interest; stochastic—the current is turned on or off

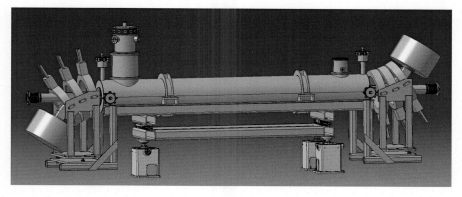

Fig. 4.10 CAD drawing of the LHC hollow electron beam collimator (courtesy S. Redaelli)

every turn according to a random function, or a random component is added on top of the constant current; and resonant—when the current is changed turn by turn according to a sinusoidal function (possibly including a frequency sweep to cover the tune spread of the proton halo), or with the same amplitude, but skipping some number of turns (as in the Tevatron abort-gap cleaning mode, see next section). The effect of the HEBC progressively grows from continuous to stochastic to resonant mode of operations.

Simulations of the HEBC at the LHC have been undertaken with the goal of estimating the magnitude of the removal rate for halo particles for realistic parameters of the low energy electron beam and high energy proton beam for different electron current variation regimes; to advise on the choice of the system geometry; and to analyze the impact of the HEBC beam imperfections on the proton beam core and the luminosity lifetime. 3D electric fields generated by a static, hollow charge distribution inside cylindrical beam pipes, including bending sections, were calculated using WARP particle-in-cell code [21, 22]. Symplectic kick maps were calculated by integrating these electric fields over straight proton trajectories [12].

Simulations show that even in the continuous mode of operation (DC electron current regime), the HEBC strongly affects the dynamics of 7 TeV protons with amplitudes larger than inner radius of the electron beam. This can be clearly seen in Fig. 4.11 from [23] that shows a frequency map analysis (FMA) of the LHC beams without (left) and with the HEBC (right). The colors progressively changing from blue to red indicate the amplitude of the betatron frequency (tune) modulation for protons starting with initial amplitudes A_x, A_y varying from $0\sigma_p$ (core) to $8\sigma_p$ (halo). One can see significantly bigger tune variations—a clear indication of enhanced diffusions in the FMA method [24, 25]—for particles with $A_{x,y} > 4\sigma_p$ in the case "HEBC on". An additional insight into the diffusion is demonstrated in Fig. 4.12

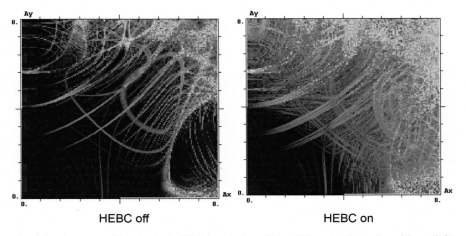

Fig. 4.11 Frequency Map Analysis (FMA) modeling of the LHC proton dynamics without (*left*) and with (*right*) HEBC in action. Horizontal and vertical axes—initial particle amplitudes A_x, A_y in units of the rms beam size varying from $0\sigma_p$ (core) to $8\sigma_p$ (halo). Brighter colors indicate exponentially stronger tune modulation indicating resonances (see text)

Fig. 4.12 Modeling of the
horizontal phase space
trajectory of LHC proton
with $5.7\sigma_p$ normalized
amplitude in case with
(*blue*) and without impact of
hollow electron beam
(*black*). The *red line*
correspond to the case of the
HEBC electron current with
random fluctuations (see
text)

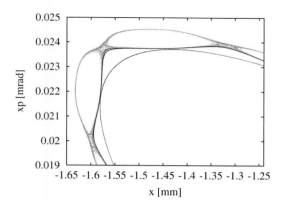

from [10] that presents a close-up of the horizontal phase space for a particle which
has a normalized amplitude of about $5.7\sigma_p$. Without the HEBC the phase trajectory
is an unperturbed ellipse (black line) which becomes strongly distorted by the
interaction with hollow DC electron beam electron lens (blue). Such distortion is
also indicative of the onset of the chaotic behavior and diffusion which can clearly
manifest itself even with minor fluctuations in the electron current—see red trajectories in Fig. 4.12. Very complex, unstable and chaotic dynamics of the LHC proton
phase-space trajectories under impact of the HEBC has been also studied with use
of the MERLIN tracking code [27].

More realistic, long-term numerical tracking simulations with the *LIFETRAC*
code [26] were used to estimate the halo removal rates for 7-TeV protons in the
nominal LHC lattice (V6.503, without magnetic multipole errors, three main
interaction points IP1, IP5, and IP8, and also at 94 long-range crossings
corresponding to the longitudinal bunch spacing of 25 ns), with nominal beam
parameters (7 TeV per beam, number of protons per bunch $N_p = 1.15 \times 10^{11}$,
transverse normalized emittance $\varepsilon = 3.75$ μm, bunch length $\sigma_z = 7.5$ cm and
momentum spread $\sigma_E = 0.00011$), with and without collisions [11]. The HEBC
element with realistic bending sections, inner electron beam radius of $4\sigma_p$ and total
electron beam current DC component variable from 0 to 3.6 A was placed at the
candidate location in IR4 where the horizontal and vertical beta-functions are equal
($\beta_x = \beta_y = 180$ m, which corresponds to the proton rms beam size of $\sigma_p = 0.32$ mm).
In the simulations, the core proton beam was represented by a 6-D Gaussian
distribution with a 6σ cut-off (Gaussian distribution in the longitudinal direction),
while 10,000 particles representing proton beam halo were uniformly distributed
between 4 and $6\sigma_p$. A single absorbing collimator was placed at $6\sigma_p$ in x and y, to
allow for the calculation of particle losses. The simulations were typically
performed on the time scale of 5×10^6 turns, which corresponds to ~450 s of the
real machine time.

Figure 4.13 presents the simulation results for HEBC in continuous mode with
total electron currents of 1.2, 2.4 and 3.6 A (the beam-beam interaction present in
three main IPs). Corresponding HEBC-induced proton halo loss rates are about

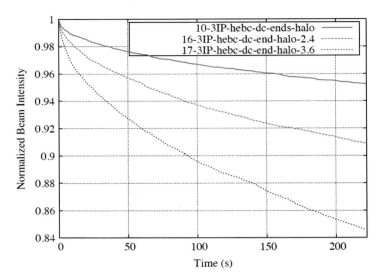

Fig. 4.13 Simulations of the LHC proton halo removal by the HEBC in continuous mode, with head-on and long-range beam-beam interactions at three collider interaction regions, and with the total electron current of 1.2 A (*red*), 2.4 A (*green dashed line*) and 3.6 A (*blue dashed line*)

28 % per hour with 1.2 A of current, 140 %/h (or 2.5 % per minute) with 2.4 A, and 250 %/h (4 %/min) with 3.6 A. It was observed that the removal rate is stronger for off-momentum particles and does not depend on the geometry of the HEBC bending sections (S-shape vs U-shape).

The stochastic mode was found to be much more robust—the introduction of stochastic turn-by-turn modulation of the electron beam current significantly enhances the halo cleaning efficiency, making the electron lens the dominant loss-driving mechanism. The cleaning rates for the cases with and without beam-beam interactions do not differ as much as in the continuous mode. In either case, 50 % of the halo is removed in 200 s by the HEBC with the maximum peak electron current of 1.2 A, while at 3.6 A the maximum cleaning rate was about 100 % per minute—see Fig. 4.14. What is critical in the stochastic mode is the HEBC system geometry, particularly—the orientation of the bending sections, where the circulating beam has to "pierce" the wall of the hollow electron beam current. The kicks from these bending sections were implemented in *LIFETRAC* code and the simulations showed no measurable effect on the LHC proton beam emittance growth and luminosity lifetime in continuous regime, in agreement with the experimental results from the Tevatron. But in the stochastic regime, the HEBC systems should be of S-shape which allows for transverse dipole kick compensation in the two bending sections and therefore, suppresses otherwise unavoidable significant emittance growth as depicted in Fig. 4.15.

Finally, the speed of the proton diffusion in the LHC beam halo can be further enhanced if the hollow electron beam current varies in sync with betatron

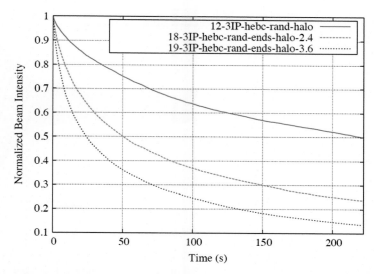

Fig. 4.14 Simulations of the LHC proton halo removal by the HEBC in stochastic mode, with head-on and long-range beam-beam interactions at three collider interaction regions, and with the total electron current of 1.2 A (*red*), 2.4 A (*green dashed line*) and 3.6 A (*blue dashed line*)

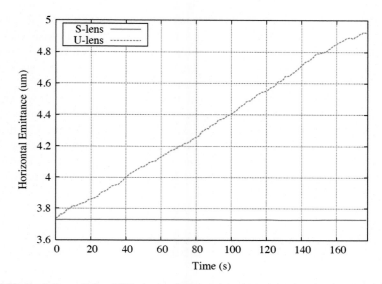

Fig. 4.15 Evolution of the LHC proton horizontal emittance under the impact a 100 % stochastically-modulated 1.2 A current hollow electron beam in two configurations: S-shape geometry (*red*) and U-shape geometry (*green*)

Fig. 4.16 The LHC proton motion driven by the hollow electron beam resonantly modulated at the tune line $Q = 0.31$ (see text)

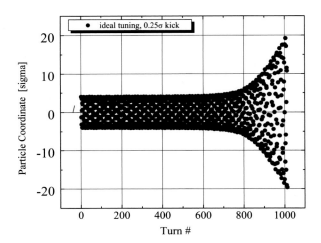

oscillations (~3 kHz AC in LHC) or at the nearest non-linear resonance line. 1D single particle simulations [5], presented in Fig. 4.16, shows a 7 TeV proton resonantly driven to amplitudes as large as 10–20σ in less than 1000 turns (0.1 s of real time in the LHC). The electron current was modulated in phase with the particle's betatron motion (tune of $Q_{betatron} = 0.31$). The amplitude (maximum strength) of the electron beam angular kick θ—see (4.1)—is equal to a quarter of the rms proton beam angular spread (for smaller or bigger kicks, the removal time scales approximately proportionally). Due to natural tune spread (induced by beam-beam, or due to synchrotron motion), one should not worry about exact synchronization of frequencies and phases with all the particles. Electron beam modulation frequency can be set close to the frequencies of interest (e.g., frequency of 4σ particles) or may cover a band of frequencies.

Figure 4.17 shows that the time needed to reach 10σ-amplitude grows with the detuning $dQ = Q_{HEBC} - Q_{betatron}$ and reaches 10 s for $dQ = 0.007$. For most optimal operation, one can envision detuning not exceeding $dQ = 0.002$ which collimates (drives particles out of the aperture set by secondary collimators) in about 0.1 s. More realistic 3D 4000-particle simulations reported in [5] confirmed that with a proper choice of Q_{HEBC} the resonant HEBC can remove halo particles in 1000–10,000 turns (0.1–1 s in the LHC).

4.2   Longitudinal Beam Collimation by Electron Lenses

Particles not captured by the machine's RF system, and, therefore, not synchronized with it, pose a significant threat to supercolliders since they can quench the superconducting magnets during acceleration or at beam abort [28]. There are several generational mechanisms of such uncaptured beam. For example, in the Tevatron at the injection stage it was coalescing in the Main Injector that typically

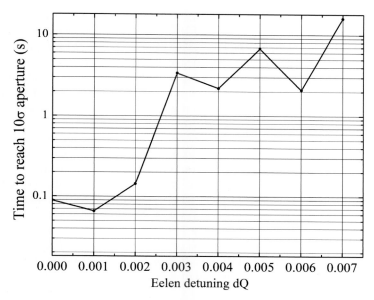

Fig. 4.17 Numerical simulation of the LHC collimation time (time needed to reach 10σ amplitude) vs. detuning parameter dQ

left a few percent of the 150 GeV beam particles outside RF buckets. These particles were transferred together with the main bunches to the Tevatron. In addition, single intra-beam scattering (the Touschek effect), diffusion due to multiple intra-beam scattering (IBS), and phase and amplitude noise of the RF voltage, drove particles out of the RF buckets. This was exacerbated by the fact that after coalescing and injection, 95 % of the particles cover almost the entire RF bucket area. The uncaptured beam was lost at the very beginning of the Tevatron energy ramp. These particles were out-of-sync with the Tevatron RF accelerating system, so they did not gain energy and quickly (<1 s) spiraled radially into the closest horizontal aperture. If the number of particles in the uncaptured beam was too large, the corresponding energy deposition results in a quench (loss of superconductivity) of the superconducting magnets and, consequently, terminations of the high-energy physics stores. At the injection energy, an instant loss of uncaptured beam equal to 3–7 % of the total intensity could lead to a quench, too, depending on the spatial distribution of the losses around the machine circumference.

At the top energy of the supercolliders, uncaptured beam generation is mostly due to the IBS and RF noise while infrequent occurrences of the longitudinal instabilities or trips of the RF power amplifiers can contribute large spills of particles to the uncaptured beam. Uncaptured beam particles are outside of the RF buckets, and therefore, move longitudinally relative to the main bunches. Contrary to the situation at the injection energy, where synchrotron radiation (SR) energy losses are practically negligible, full energy protons or antiprotons lose noticeable energy due to the SR, e.g., about 9 eV/turn in the Tevatron and some 560 eV/turn for 7 TeV particles in the LHC. For uncaptured beam particles, this energy loss is not being replenished by the RF system, so they slowly spiral radially inward and die on the collimators, which usually determine the tightest aperture in the colliders. The typical time for an uncaptured particle to reach the collimator is about 20 min in the Tevatron and a few minutes in the LHC. Detail theoretical analysis of various mechanisms of the uncaptured beam generation can be found in [29].

The presence of the uncaptured beam could be very dangerous not only for the collider elements but also for the high-energy physics particle detectors, as it was for the CDF and D0 experiments in the Tevatron, as the abort gap particles generate unwanted background and can be kicked onto the detectors' components by the beam abort kickers. A number of ideas had been proposed for elimination of the uncaptured beam in the Tevatron. The Tevatron electron lenses were found to be the most effective. The advantages of the TELs were twofold: (i) an electron beam can be placed in close proximity to proton or antiproton orbits and generates a very strong transverse kick; (ii) the electron current pulses have short rise and fall times (~100 ns), so they can be easily adjusted to operate in a variety of different pulsing schemes.

The first Tevatron electron lense was installed in the Tevatron in 2001 and alraedy in early 2002 it was found to be an effective remover of uncaptured protons if timed into the abort gap and operated in a resonant excitation regime [28]. TEL-2 was also able to function as an abort gap "cleaner". For that, the electron beam pulse was synchronized to the abort gap and positioned near the proton beam orbit. Electric and magnetic forces due to the electron space charge produce a radial kick on high-energy (anti)protons depending on the separation d:

$$\Delta\theta = \mp \frac{1 \pm \beta_e}{\beta_e} \cdot \frac{2 J_e L_e r_p}{e \cdot c \cdot \gamma_p} \cdot \begin{cases} \dfrac{d}{a}, & d < a \\ \dfrac{a}{d}, & d > a \end{cases} \tag{4.7}$$

where the sign reflects repulsion for antiprotons and attraction for protons, $\beta_e = v_e/c$ is the electron beam velocity, J_e and L_e are the electron beam current and the interaction length, a is the electron beam radius, r_p is the classical proton radius, and γ_p is the relativistic Lorentz factor (1044 for 980 GeV (anti)protons). The factor $1 \pm \beta_e$ reflects the fact that the contribution of the magnetic force is β_e times the electric force contribution and depends on the direction of the electron velocity.

For typical Tevatron electron lens parameters—5 kV electrons with peak current of about 0.6 A, placed some 5 mm away from the protons—the estimated kick was about $\Delta\theta = 0.07$ μrad. When the pulsing frequency of the TEL was near the proton beam resonant frequency, this beam-beam kick resonantly excites the betatron oscillations of the beam particles.

In the uncaptured beam removal operation, the TEL electron beam was placed 2–3 mm away from the proton beam orbit horizontally and about 1 mm down vertically as depicted in Fig. 4.18. The Tevatron horizontal and vertical tunes of

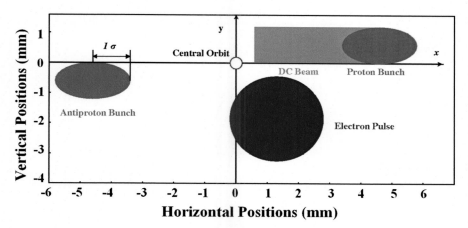

Fig. 4.18 The relative positions of the proton, antiproton and the TEL beam for the uncaptured beam removal in the Tevatron [28]

$Q_x = 0.583$ and $Q_y = 0.579$ (fractional parts), respectively, were between the strong resonances at $4/7 \approx 0.5714$ and $3/5 = 0.6$. When an uncaptured particle loses energy due to synchrotron radiation, its horizontal orbit shifts proportionally to the lattice dispersion $x = D_x(dP/P)$ and the betatron tunes slide due to the lattice chromaticity $C_{x,y} = dQ_{x,y}/(dP/P)$:

$$Q_{x,y} = Q^0_{x,y} + C_{x,y}\left(\frac{dp}{p_0}\right) + \Delta Q_{x,y}(x^2), \tag{4.8}$$

where the third term accounts for slight tune changes due to nonlinear magnetic fields. Typical chromaticities of the Tevatron at 980 GeV were set to $C_{x,y} \approx +(6-10)$, so the tune decreased with the energy loss. As the TEL current modulation frequency is set close to one of the resonant lines, the amplitude of the particle's betatron oscillations grows, eventually exceeding a few millimeters until the particle gets intercepted at the collimators. The maximum amplitude is determined by the nonlinearity of the force due to the electron beam and the nonlinearity of the machine. Note that without the TEL, a particle would still be intercepted by a horizontal collimator after its orbit moved about 3 mm inward due to the synchrotron radiation enery loss. The TEL simply drives the particles onto collimators more quickly, preventing the accumulation of uncaptured beam.

The electron beam pulsing scheme is demonstrated in Fig. 4.19, where the green oscilloscope trace is the signal from the TEL Beam Position Monitor (BPM) pickup electrode and the blue trace is the total electron current. In the BPM signal, one can see three negative pulses representing the electron beam pulses in the three abort gaps whereas the 36 positive pulses are the proton bunch signals with the small negative adjacent antiproton bunch signals. The intensity of the antiproton bunches was ten times less than that of the proton bunches at the end of that particular store,

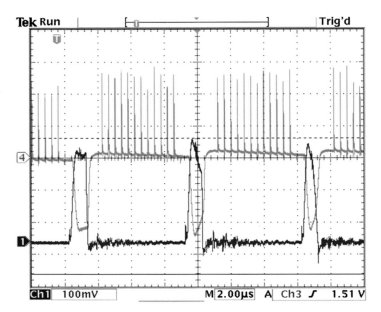

Fig. 4.19 Scope traces of the electron beam pulses (*blue*) and the TEL BPM signal showing electron, proton and antiproton bunches (*green*). One division of the horizontal axis is 2 μs. About one Tevatron revolution period is shown [28]

so they appear only as very small spikes near the large proton bunches. During a typical HEP store, the train of three electron pulses is generated every seventh turn to excite the 4/7 resonance and achieve effective removal of the uncaptured proton beam particles. The electron pulse width is about 1 μs and the peak current amplitude is about 250 mA.

Efficiency of the uncaptured beam removal process was demonstrated in an experiment in which the TEL was turned off for about 40 min and then turned on again as shown in Fig. 4.20. The blue trace is the total bunched proton beam intensity measured by the Fast Bunch Integrator [30]; the red trace is the average electron current measured at the TEL electron collector; the green trace is the total number of particles in the Tevatron as measured by DC beam current transformer [30]; and the cyan trace is the abort gap proton beam loss rate measured by the CDF detector counters.

After the TEL was turned off, the abort gap loss rate was reduced by about 20 % but then started to grow. After about 20 min, the first spikes of the proton losses started to appear and grow higher. Notably, the bunched beam intensity (blue line) loss rate did not change, so the rate of particles escaping from the RF buckets was about constant. As soon as the TEL was turned on, a large increase in the abort gap losses and reduction of the total uncaptured beam intensity could be seen. About 15×10^9 particles of the uncaptured beam in the abort gaps were removed by the TEL in about $\tau_{TEL} = 3$ min and the abort gap loss rate went back to a smooth equilibrium baseline.

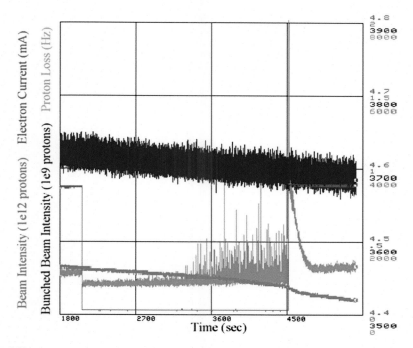

Fig. 4.20 Uncaptured beam accumulation and removal by TEL: the electron current was turned off and turned back on 40 min later again (see text)

The calibration of the abort gap monitor (AGM, used for the routine monitoring of the uncaptured beam—see details in [31]) had been performed using the TEL as presented in Fig. 4.21. The TEL was turned off during a store (average electron current is shown in black) at about $t = 20$ min. Accumulation of the uncaptured beam started immediately and can be measured as an excess of the total uncaptured beam current with respect to its usual decay. The blue line in Fig. 4.21 shows the excess measured by the Tevatron DCCT, $\delta N_{DCCT}(t) = N_{TEL\ on}(t) - N_{decay\ fit\ TEL\ off}(t)$. The abort gap uncaptured beam intensity measured by the AGM (red line) and the DCCT excess grew for about 30 min before reaching saturation at intensity of about 16×10^9 protons. Then the TEL was turned on resulting in the quick removal of the accumulated uncaptured beam from the abort gaps. The amount of the uncaptured beam is determined by the rate of its generation and the removal time τ_{TEL}:

$$N_{DC} = \left(\frac{dN_{bunched}}{dt}\right) \times \tau_{TEL} \tag{4.9}$$

The characteristic time needed for a 980 GeV particle to lose enough energy due to the synchrotron radiation is about $\tau_{SR} = 20$ min, so the TEL reduces the uncaptured beam population by about τ_{TEL}/τ_{SR}, i.e., by about an order of magnitude.

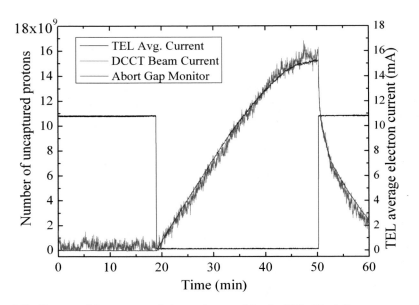

Fig. 4.21 Uncaptured beam accumulation and removal by the TEL. *Black line* represents the average electron current of the TEL; *red line* is the uncaptured beam estimated from the DCCT measurement; *blue line* is uncaptured beam in the abort gap measured by the Tevatron Abort Gap Monitor [31]

At injection energy, the synchrotron radiation of protons is negligible, so the TEL was the only means to control uncaptured beam. As noted above, one of the TELs is used routinely in the Tevatron operation for the purpose of uncaptured beam removal at 150 and 980 GeV. In 2007, the typical antiproton intensity increased to about a third of the proton intensity, and therefore the antiproton uncaptured beam accumulation started to pose an operational threat. An antiproton AGM, similar to the proton one, was built and installed. With proper placement of the TEL electron beam between the proton beam and the antiproton beam (illustrated in Fig. 4.19), it was possible to remove effectively both uncaptured protons and uncaptured antiprotons. The effectiveness of the uncaptured beam removal was explored at several resonant excitation frequencies. For that, the TEL had been pulsed every second, third, fourth, fifth, sixth and seventh turn. Reduction of the uncaptured beam intensity was observed at all of them, though usually the most effective was the every seventh turn pulsing when the Tevatron betatron tunes were close (slightly above) to $Q_{x,y} = 4/7 = 0.571$ or every sixth turn pulsing when tunes were closer to $Q_{x,y} = 7/12 = 0.583$.

To conclude, the Tevatron experience confirmed very high efficiency of the uncaptured beam collimation by the electron lenses, i.e., the longitudinal beam collimation. In a similar fashion, the space charge forces of electron lenses can be used for selective elimination—randomly or via resonant excitation—of unwanted particles or bunches, e.g., in satellite RF buckets.

References

1. V. Shiltsev, in *Proceedings of the 3rd CARE-HHH-APD Workshop (LHC-LUMI-06)* (Valencia, 2007), p. 92, CERN-2007-002
2. V. Shiltsev, in *Proceedings of the CARE-HHH-APD Workshop (BEAM07)* (Geneva, 2008), p. 46, CERN-2008-005
3. G. Stancari et al., Phys. Rev. Lett. **107**, 084802 (2011)
4. G. Stancari et al., arxiv:1405.2033; also as CERN-ACC-2014-0248 FERMILAB-TM-2572-APC (2014)
5. V. Shiltsev et al., in *Proceedings of 2008 EPAC* (Genoa, Italy, 2008), p. 292
6. K.H. Mess, M. Seidel, Nucl. Instr. Meth. Phys. Res. A **351**, 279–285 (1994)
7. G. Stancari, in *Proceedings of the Meeting of the Division of Particles and Fields of the American Physical Society* (Providence, August 2011); arXiv:1110.0144 (2011)
8. G. Stancari et al., in *Proceedings of the 52nd ICFA Advanced Beam Dynamics Workshop on High-Intensity and High-Brightness Hadron Beams (HB2012)* (Beijing, September 2012), p. 466, FERMILAB-CONF-12-506-AD-APC
9. See Chapter 2.5 in the book V. Lebedev, V. Shiltsev, *Accelerator Physics at the Tevatron Collider* (Springer, New York, 2014), and references therein.
10. V. Previtali et al., FERMILAB-TM-2560-APC (July 2013)
11. A. Valishev, FERMILAB-TM-2584-APC (May 2014)
12. G. Stancari, FERMILAB-FN-0972-APC, arXiv:1403.6370 (2014)
13. R.W. Aßmann et al., in *Proceedings of EPAC'2002* (Paris, France, 2002), pp. 197–199
14. R.W. Aßmann et al., in *Proceedings of EPAC'2006* (Edinburgh, UK, 2006), pp. 986–988
15. G. Valentino, R.W. Aßmann, R. Bruce, S. Redaelli, A. Rossi, N. Sammut, D. Wollmann, Phys. Rev. ST Accel. Beams **15**, 051002 (2012)
16. B. Salvachua et al., in *Proceedings of IPAC'13* (Shanghai, China, 2013), pp. 1002–1004
17. L. Rossi, in *Proceedings of IPAC'11* (San Sebastian, Spain, 2011), pp. 908–910
18. S. Li, G. Stancari, FERMILAB-TM-2542-APC (2012)
19. V. Moens, *Master Thesis*, École Polytechnique Fédérale de Lausanne (EPFL), Switzerland, FERMILAB-MASTERS-2013-02 and CERN-THESIS-2013-126 (2013)
20. S. Redaelli et al., in *Proceedings of IPAC'2015* (Richmond, VA, USA, 2015), p. 2462
21. See, e.g., A. Friedman, D.P. Grote, I. Haber, Phys. Fluids B **4** (1992), p. 2203
22. Q. Ji et al., in *Proceedings of the 52nd ICFA Advanced Beam Dynamics Workshop on High-Intensity and High-Brightness Hadron Beams (HB2012)* (Beijing, 2012), p. 546
23. G. Stancari, 'Beam experience at the Tevatron and status of the hollow electron-lens hardware,' Talk presented at the *Special Collimation Upgrade Specification Meeting: Internal Review of Tevatron Hollow Electron-Lens Usage at CERN*, https://indico.cern.ch/event/213752 (Geneva, 9 November 2012)
24. J. Laskar, Icarus **88** (1990), pp. 266–291
25. J. Laskar, in *Proceedings of 2003 IEEE PAC* (Portland, OR, USA, 2003), p. 378
26. D. Shatilov et al., in *Proceedings of 2005 IEEE PAC* (Knoxville, TN, USA, 2005), p. 4138
27. H. Rafique et al., in *Proceedings of IPAC'2015* (Richmond, VA, USA, 2015), p. 2188
28. X.-L. Zhang et al., Phys. Rev. ST Accel. Beams **11**, 051002 (2008)
29. See Chapter 6.5 in the book V. Lebedev, V. Shiltsev, *Accelerator Physics at the Tevatron Collider* (Springer, New York, 2014), and references therein
30. R.S. Moore, A. Jansson, V. Shiltsev, JINST **4**, P12018 (2009)
31. N. Mokhov et al., JINST **6**, T08005 (2011)

Chapter 5
Electron Lenses for Space-Charge Compensation, Other Applications of Electron Lenses

Unique properties and flexibility of the electron lenses allow many other important accelerator applications beyond the head-on and long-range beam-beam compensation and the transverse and longitudinal hadron beam halo collimation. Below we consider several of the most actively studied and pursued ideas and proposals, including the electron-lens compensation of space-charge effects in high-intensity proton accelerators, including super-collider injectors; attainment of nonlinear integrable beam dynamics to suppress halo formation and particle loss in high-brightness proton rings; selective bunch-by-bunch or batch-by-batch slow extraction systems; beam-beam compensation in $e + e-$ colliders and electron-ion colliders, tune-spread generators for Landau damping of coherent beam instabilities, and the "beam-beam kicker" .

5.1 Space Charge Compensation with Electron Lenses

5.1.1 Electron Lens Compensation of Space-Charge Effects: Theory and Modeling

The term "space charge effects" covers a number of important phenomena in accelerators originating from self-fields of the charged particle beams. The forces in a parallel flow of charges include repelling force due to self electric field $E(r)$ and attractive force due to self magnetic field which is smaller than the former by a factor proportional to particles' velocity $\beta = v/c$. The resulting net force is repelling (defocusing), it usually varies across the beam and scales with the charge density n_p and relativistic factors as $F(r) = e(E(r) - \beta B(r)) = eE(r) \times (1 - \beta^2) \sim n_p/\gamma^2$. Depending on the strengths of these fields and the time scales, the consequences may include very fast beam blowup and losses in particle sources or single-pass systems (linacs), dangerous emittance growth, beam-brightness degradation and

© Springer Science+Business Media New York 2016
V.D. Shiltsev, *Electron Lenses for Super-Colliders*, Particle Acceleration and Detection, DOI 10.1007/978-1-4939-3317-4_5

performance-limiting beam halo development, instabilities and losses in the systems operating over hundreds or thousands of turns (cyclotrons, rapid cycling synchrotrons), and lifetime reduction and complex non-linear beam dynamics in storage rings and colliders operating over many millions of turns. The space charge effects have been well known in accelerator physics for a long time and there is extensive literature on the corresponding theory and modeling [1–15]; they are covered in a textbook [16] and in a concise review in the *Accelerator Handbook* [17]. Comprehensive review of the phenomenology of the space charge effects in proton accelerators can be found in [18], recent papers overview the space-charge related challenges in the LHC injectors [19], electron ion-colliders [20, 21], and high intensity accelerators for neutrino physics experiments [22, 23].

In circular accelerators, the nonlinear space charge forces induce an irreducible transverse tune spread, e.g., the tune dependence on both the particle's longitudinal position z inside the bunch and the amplitude of the transverse betatron oscillations a:

$$\Delta Q_{SC}(a, z) \approx \Delta Q_{SC} \cdot \frac{I_0\left(a^2/4\sigma_r^2\right)}{\left(a^2/4\sigma_r^2\right)} \left[1 - \exp\left(a^2/4\sigma_r^2\right)\right] \cdot \exp\left(-\frac{z^2}{2\sigma_z^2}\right),$$

$$\Delta Q_{SC} = -\frac{Z^2 r_p}{A} \frac{N_p}{4\pi\varepsilon_n\beta_p\gamma_p^2\sigma_z\sqrt{2\pi}}, \tag{5.1}$$

where $I_0(x)$ is the modified Bessel function of order 0, and the estimate is given for a ring with circumference C, Z and A are the particle's charge and atomic number (e.g., for protons $Z = A = 1$), N_p is number of particles in the axisymmetric Gaussian bunch with rms length of σ_z and rms transverse size of σ_r. Thus, the tune shift for particles in the beam core and the bunch center is larger than for those in the transverse or longitudinal tails. Given that the relativistic factors β_p, γ_p for hadrons (protons, ions) are usually lower than for electrons, the space charge effects set very stringent limits on the maximum operationally achievable beam brightness and power: the space-charge limit is about $|\Delta Q_{SC}| \approx 0.2$–$0.4$ in rapid cycling synchrotrons and $|\Delta Q_{SC}| \approx 0.05$–$0.1$ in storage rings and colliders.

A number of schemes for compensating space charge effects in hadron beam were proposed and some of them have been tested [24]. Passive cancellation of the next-to-leading term in the space charge force is possible by octupole fields. Indeed, for a round beam, the space charge potential $\varphi_{SC}(r)$ and field $E_{SC}(r)$ are related to the charge distribution $n(r)$:

$$\Delta\varphi_{SC}(r) = -4\pi n(r), \quad \vec{E}_{SC} = -\vec{\nabla}\varphi_{SC}, \tag{5.2}$$

and therefore, the potential can be presented as a series of powers of $r^2 = (x^2 + y^2)$. The second term results in the radially defocusing linear force and in principle can be corrected by the machine tune correction circuits. The next order term of the direct space charge potential varies as $r^2 = (x^4 + 2x^2y^2 + y^4)$, while the potential of a single octupole magnet is proportional to $(x^4 - 6x^2y^2 + y^4)$. Therefore, at least two families of octupoles are needed to reduce the space-charge induced tune spread, which are placed at locations with either peak or intermediate values of the beta function, respectively. The beta functions should sufficiently vary over the length of an optical cell, e.g., by a factor 2 or more. Magnet pole-face windings could in principle allow more precise adjustments of the tune shift with transverse position up to a higher order. At the CERN Intersecting Storage Rings (ISR), 24 pole-face windings modifying the local magnetic field were used to correct the horizontal and vertical indirect space charge tune shift plus the next four orders in their Taylor expansions with respect to the horizontal position x. The correction increased the maximum ISR beam current by more than an order of magnitude [25]. The major difficulty in this approach is an intrinsic difficulty mimicking fields of the non-Laplacian space charge potential $\Delta\varphi_{SC} \neq 0$ by those of the Laplacian potentials of electromagnets $\Delta\varphi_{MAG} = 0$. The later represented as series of $x^m y^n$, where m,n are integer and, for example, an infinite number of the diverging terms are needed to approximate simple space-charge field scaling at large radii $E_{SC}(r) \sim 1/r = 1/(x^2 + y^2)^{1/2}$.

A more promising approach is to compensate positive scape charge of the proton beam by negative space charge of stationary or low-energy electrons as illustrated in Fig. 5.1. If the charge profiles of protons and electrons are functionally the same, e.g., Gaussian, the compensation requires only a relatively small number of electrons $\eta = N_e/N_p \approx 1/\gamma_p^2$. In the case of passive neutralization, the space charge force of a proton beam can compensated by ionization electrons, electron cloud, or negative ions which are approximately at rest longitudinally, but move transversely during the beam passage. Neutralized low energy beams of heavy ions have successfully transported in a number of linear accelerators [26]. Almost an order of magnitude increase of the maximum circulating beam current above (coherent) space charge limit was achieved at the Novosibirsk 1 MeV proton ring by increasing the residual gas pressure in excess of 10^{-4} Torr and accumulation of ionization electrons [27]. Unfortunately, the observed beam lifetime was very short and

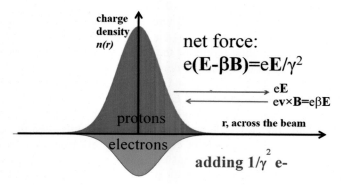

Fig. 5.1 Illustration of the space-charge effects in proton beams and the principle of electron beam compensation (see text)

transverse and longitudinal proton and electron distributions were not well controlled.

Electron lenses in which externally generated electron beam with matched transverse distribution collides with the proton beam inside a strong solenoid field, were proposed for effective and controlled compensation of the space-charge effects in high intensity proton rings [28]. Protons going through the electron beam experience focusing force which has an opposite sign to the proton beam own space charge force and can precisely fully or partially compensate the latter if:

1. Transverse profile of the electron beam charge $n_e(r)$ is the same as proton beam profile,
2. Integrated impact of electrons is equal to the total proton space charge impact over the ring,
3. Temporal structure of the electron space charge force matches the proton space charge force.

The first condition can be satisfied without difficulties: the electron beam profile can be controlled by special electrodes in the electron gun and via magnetic compression to match the proton profile and the rms size $\sigma^2_{e,r} = \sigma^2_{p,r} = \varepsilon_p \beta_p \gamma_p \beta_r$ the same way as in the electron lenses for beam-beam compensation, e.g., the TEL—see above in Sect. 2. Transverse rigidity of the electron beam in the solenoidal magnetic field in the interaction region effectively prevents the low energy electron profile distortions under the impact of protons. In practice, the electron beam set-ups could occupy only a small fraction of the ring circumference C. Assuming the total length of all electron lenses $L = N_{EL}L_{EL}$, the condition of (partial) compensation on average reads $\Delta Q_{EL} = -\Delta Q_{SC} \times \kappa$—see (5.1), (1.17), (3.1), κ being a degree of the compensation—yielding the electron current requirement in each lens [28]:

$$J_e = \frac{B_f ecN_p}{L_{EC}} \frac{\beta_e}{\gamma_p^2 \beta_p^2 (1 - \beta_e \beta_p)} \kappa, \qquad (5.3)$$

where B_f is the proton bunching factor, defined as the ratio of the maximum to average current within one RF bucket. There are two factors which determine the number of required electron lenses N_{EL}—one is that shorter electron lenses allow for a better match of the time profile of the electron impact to that of the proton bunch current, another is that a higher degree of periodicity of the ring focusing lattice helps to improve single particle and collective stability. Both effects call for a larger number N_{EL} and shorter length L_{EL}.

Indeed, the necessity of fast longitudinal electron current modulation for matching the proton bunch profile and space-charge tuneshift—see the last factor in (5.1)—calls the relative slippage of co-propagating electrons and protons over the interaction length L_{EL} not to exceed by far the proton bunch length σ_z, i.e., $L_{EL} \sim F \sigma_z \beta_e/(\beta_p - \beta_e)$ (here F is a numerical factor of the order of 1). Out of this consideration, it is definitely advantageous to have the electron and proton beams moving in the same direction, shorter L_{EL} and the required proton current pulse

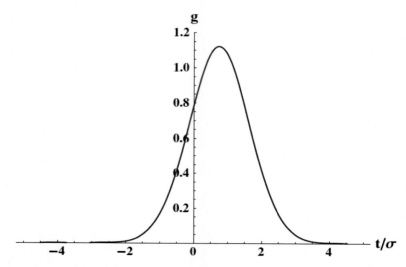

Fig. 5.2 The electron waveform needed for perfect space-charge compensation for electron beam system length equivalent to the slippage factor of $F = 1.5$, the time axis is in the units of the rms bunch length σ_z/c of the Gaussian proton bunch (see text) [29]

current waveform can be computed using deconvolution method proposed in [29]—see Fig. 5.2. An alternative approach could be to flatten the proton bunch longitudinal distribution using, e.g., a higher harmonics RF systems, so a DC electron current can effectively compensate the space charge effects for essentially all protons.

Given that the space-charge tuneshift to compensate is usually large $-\Delta Q_{SC} \sim 0.2$–0.5 or more, having just one strong electron lens may not be feasible because of the very high current requirement (5.3) and an unacceptably large disruption of the machine's focusing lattice. That would make the focusing lattice periodicity P equal to 1, and would lead to very fast proton beam loss due to enhanced strength of incoherent and coherent resonances [30].

Symmetry of an accelerator focusing lattice is important for the space-charge effects and compensation. If the lattice consists of P identical periods, then strong and wide *structure* linear resonances occur at coherent/incoherent frequencies $Q_{coh/incoh} = Pm/2$, where m is integer. All other integer and half-integer tunes relate to relatively weak and narrow *non-structure* linear resonances, excited by periodicity perturbations (errors). Having high machine periodicity and a properly chosen working point (Q_x, Q_y), one can have all the nearest linear incoherent and coherent resonances to be non-structure resonance, and thus allow operation with larger space-charge tune spread ΔQ_{SC}. Maintaining high periodicity is critical for the space charge compensation as well. For example, numerical simulations for the Fermilab Booster ring with periodicity $P = 24$, $N_p = 6 \times 10^{10}$ and tunes $Q_{x,y} = (6.7, 6.8)$ showed drastic reduction of the emittance blowup and beam losses if electron beam compensating devices were placed in each of the 24 Booster lattice

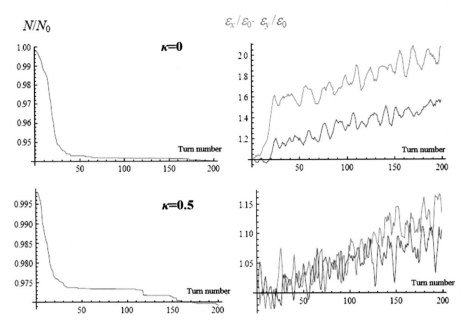

Fig. 5.3 Fermilab Booster ring simulation over 200 turns without space-charge compensation (*top*) and with partial compensation $\kappa = 0.5$ by 24 localized electron devices (from [31]). On the *left*—normalized proton intensity loss; on the *right*—growth of vertical and horizontal emittances normalized to their initial values. These 2D simulations have been carried out with $N_p = 6 \times 10^{10}$ and $-\Delta Q_{SC} \sim 0.4$

periods [31, 32]—see Fig. 5.3. With fewer compensators $N_{EL} = 12$, the beneficial effect of compensation is somewhat smaller for the same values of the compensation factor κ. It is of note that even partial compensation $\kappa = 0.5$ results in significant reduction of the beam losses. Table 5.1 presents parameters of the electron lenses for the space charge compensation in the FNAL Booster.

Similar simulations of the beam dynamics with $\Delta Q_{SC} = -0.2$ over the first 256 turns in the KEK Proton Synchrotron ($C = 340$ m, $E_{inj} = 500$ MeV, $P = 4$) show that the electron beam matched to the proton distribution effectively suppresses the emittance growth due to space charge effects at $\kappa = 0.5$–1.0 even with relative electron-proton orbit misalignments of about $(0.1$–$0.2)\sigma_p$ [33]. Another numerical tracking study of the electron lens compensation in the CERN PS Booster ($C = 157$ m, $E_{inj} = 50$ MeV, $P = 16$) concluded that four electron lenses can effectively reduce the initial space charge tune spread by about $\Delta Q_{SC} = -0.5$, that higher number of lenses (e.g., 8) should be beneficial, and the electron current modulation matching the proton bunch current profile is important to avoid overcompensation in the longitudinal tails of proton bunches [19]. As for the concerns expressed in [28] that degree of the compensation higher than $\kappa = 0.33$ might result in excitation of coherent space-charge modes, it was pointed out that such modes have not been clearly observed in the low-energy CERN proton accelerators or in the simulations.

Table 5.1 Main parameters of the electron lenses for the space-charge compensation in the Fermilab's Booster rapid cycling proton synchrotron

Parameter	Symbol	Value			Unit
Electron lens parameters, κ—degree of compensation					
Number of e-lenses	N_{EL}	24	12	6	
Length of each e-lens	L_{EL}	0.7	1.0	2.0	m
Peak *e*-current (max)	J_e/κ	2.0	3.6	4.2	A
e-beam energy (max)	U_e	30	40	50	kV
Magnetic field in main solenoid	B_m	1.1			T
Magnetic field in gun solenoid	B_g	0.3			T
e-beam radius in main solenoid	σ_e	4.5			mm
Cathode radius	a_c	12			mm
Total tuneshift by e-lenses (max)	ΔQ_{EL}	0.4			×κ
Booster synchrotron parameters					
Circumference	C	474			m
Lattice periodicity	P	24			
Proton energy, injection/extraction	E_{kin}	0.4/8.0			GeV
Cycle time	T_c	67			ms
p-bunch intensity	N_p	~60			10^9
Number of bunches	N_B	81			
Emittance (normalized, rms)	ε_p	≈1–3			μm
Rms bunchlength at injection	σ_z	≈1			m
Maximum space-charge tune shift	$-\Delta Q_{SC}$	~0.4			

5.1.2   Space-Charge Compensation Experiments at IOTA ring

Progress of the intensity frontier accelerator-based high energy particle physics programs, such as neutrino physics and rare decays, is hindered by fundamental beam phenomena such as space-charge effects and their compensation, beam halo formation, particle losses, transverse and longitudinal instabilities, beam loading, inefficiencies of beam injection and extraction, etc. [23]. The Integrable Optics Test Accelerator (IOTA) facility [34–36]—see Fig. 5.4—is being built at Fermilab as a unique test-bed for transformational R&D towards the next generation high-intensity proton facilities. The experimental accelerator R&D at the IOTA ring with protons and electrons, augmented with corresponding modeling and design efforts will lay a foundation for novel design concepts, which will allow substantial increase of the proton flux available for HEP research with Fermilab accelerators to multi-MW beam power levels at affordable costs. Experimental demonstration of the compensation of the space-charge effects with electron lenses is one of the key goals of the IOTA program.

The IOTA facility—see parameters in Table 5.2—consists of a 40 m circumference storage ring and injectors of electrons and protons. The ring is unique as it can operate with either narrow "pencil-like" electron beams (up to 150 MeV/c

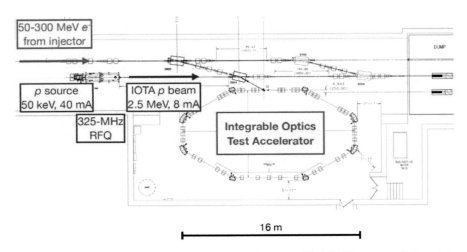

Fig. 5.4 Layout of the Integrable Optics Test Accelerator (IOTA) facility at Fermilab, and its main components: 50–300 MeV electron injector, 2.5 MeV proton injector consisting of the proton source and 325 MHz RFQ accelerator and the IOTA storage ring

Table 5.2 Main parameters of IOTA proton ring and its electron lens for the space-charge compensation [37, 38]

Parameter	Symbol	Value	Unit
IOTA Ring with Proton Injector			
Circumference	C	39.97	M
Proton kinetic energy	E_{kin}	2.5	MeV
Proton momentum	P	68.5	MeV/c
Proton velocity	$\beta_p = v_p/c$	0.073	
Total number of protons	N_p	9.1	10^{10}
Emittance (normalized, rms)	ε_p	0.3	mm mrad
Maximum space-charge tune shift	$-\Delta Q_{SC}$	~0.5	$\times B_f$
Beta-function at the e-lens	β_x	3.0	m
Electron Lens at IOTA			
Length	L_{EL}	0.7	m
Current (max, DC)	J_e	1.1–3.0	A
e-beam energy (max)	U_e	5–10	kV
Magnetic field in main solenoid	B_m	0.3	T
Magnetic field in gun solenoid	B_g	0.1	T
e-beam radius in main solenoid	σ_e	2	mm
Current density on axis	j_e	9.0	A/cm^2
Cathode radius	a_c	12–22	mm
Tune shift by e-lens (max)	ΔQ_{EL}	0.15–0.25	

momentum) free of space change effects or with heavily space-charge dominated 70 MeV/c proton beams from a 325 MHz RFQ injector [37]. The ring has a large aperture, significant flexibility of the focusing lattice, the means for precise control of the optics quality and all the diagnostics needed for experimental studies with "pencil" (narrow) electron beams and very high intensity proton beams. A 0.7-m long electron lens in the S-shape configuration with few Amperes of the low energy electron beam current will be installed in one of the IOTA straight sections—see Fig. 5.5. The goal of the corresponding experimental studies is to demonstrate full or partial compensation of the space-charge effects and stable particle dynamics in the high-intensity proton beams with $\Delta Q_{SC} = -\kappa \times \Delta Q_{EL} \sim 0.25$–$0.5$ by using Gaussian transverse current distribution $j_e(r)$ in the electron lens. The IOTA experiment configuration is an ideal approximation of a realistic accelerator design (magnets, injection system, RF system, etc.) and could represent a single-cell model of the ultimate multi-MW proton rapid cycling synchrotron with record high total space-charge parameter $\Delta Q_{SC} = -(\kappa \times \Delta Q_{EL}/\text{cell}) \times N_{cells}$.

Besides the compensation studies with the Gaussian electron lens, synergetic explorations of the *integrable* electron lenses and "electron columns" are planned. The concept of nonlinear integrable optics applied to accelerators involves a small number of special nonlinear focusing elements added to the lattice of a conventional machine in order to generate large tune spreads while preserving dynamic aperture [39, 40]. The concept promises improved stability to perturbations and mitigation of collective instabilities through Landau damping. The integrability of

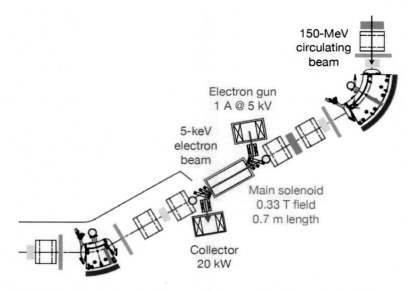

Fig. 5.5 Layout of the electron lens for space-charge compensations studies in IOTA

axially symmetric thin-lens kicks was studied in one dimension by McMillan [41, 42]. It was then extended to two dimensions [43] and experimentally tested with colliding beams [44]. In [39] it was pointed out that a linear accelerator lattice possesses additional integrals of particle motion if it includes electron lens with a specific radial dependence of the current density j_e (r):

$$j_e(r) = \frac{J_e}{\pi a_e^2} \frac{1}{\left(1 + \frac{r^2}{a_e^2}\right)^2},$$ (5.4)

where J_e is total electron current, and a_e is effective beam radius. Beam of protons circulating in such accelerators experiences nonlinear transverse kicks due to the lens which scale as:

$$\theta(r) \propto \frac{r}{1 + \frac{r^2}{a_e^2}}.$$ (5.5)

For such a radial dependence of the kick, and if the lens element is thin ($L_e \ll \beta_{x,y}$), there are two independent invariants of motion in the four-dimensional transverse phase space. Neglecting longitudinal effects, all particle trajectories are regular and bounded. The achievable nonlinear tune spread ΔQ, i.e., the tune difference between small and large amplitude particles, is about ΔQ_{EL}. Generation of the required distribution (5.4) in the electron gun of the IOTA electron lens is straightforward and corresponding experiments are being planned [37].

The concept of the space-charge compensation by *electron columns* [45] is very similar to the one with electron lenses but assumes passive neutralization of the proton space charge by properly accumulated and stored electrons created via the acts of ionization of the residual vacuum molecules by the proton beam itself. The method does not need an electron gun and collector, but to be effective it requires both protons and electrons to be immersed in a strong longitudinal magnetic field to keep electrons from escaping from the transverse position in which they are born and strong enough to assure transverse multi-turn stability of the electron-proton system. The process of electron accumulation and charge compensation also needs efficient removal of the ions from the electron column section. Analytical and numerical studies of the electron column demonstration experiment in IOTA have been started [46].

5.2 Electron Lenses for Slow Extraction from Proton Synchrotrons

There are a number of ways to extract beams out of circular accelerators [47, 48]. The most generally used in proton synchrotrons are single-turn extraction using fast kickers and slow resonant multi-turn extraction. The resonant multi-turn extraction allows delivery of beam to experiments over a time interval from milliseconds to hours. Non-linear fields of slow bumper magnets excite betatron resonances which drive the beam slowly across the extraction septum—these are often third-order resonances $3Q_{x,y} = n$ excited by sextupole magnets or second order linear resonances $2Q_{x,y} = m$ in combination with octupole magnets (n,m are integer). The nonlinear fields distort the circular normalized-phase-space particle trajectories so a stable area is controlled by the tune-to-resonance distance and strength of the nonlinearities. Flexibility of the electron lenses offers new opportunities for the multi-turn slow extraction method. Negatively charged low-energy electron beam overlapping with protons results in positive proton tune shift ΔQ_{EL} proportional to the electron current—see (1.17). If the resulting tune satisfies the non-linear resonance condition, protons move to larger betatron amplitudes as in the conventional slow extraction systems. Time variation of the electron current allows slow extraction from particular batches or even slow extraction from a single bunch.

Let us illustrate the use of an electron lens for slow extraction of proton bunches from the Fermilab Main Injector. The Main Injector is a 3.3 km long 120 GeV proton synchrotron designed to provide beams to many areas, including the Tevatron collider, antiproton production, fixed target experiments [49]. It is possible to extract the beam from a portion of the Main Injector (MI) circumference using fast (single turn)extraction kickers and then extract slowly the remaining beam with resonant extraction system [50]. In some scenarios it is useful to invert the order, namely, to use slow extraction on a portion of the circumference while retaining some of the beam for later extraction. This scenario can become a reality if a large bandwidth proton tune shifter (with risetime much less than a turn) is available. An electron lens provides a suitable technology for this application [51].

The standard way of the resonant extraction in the MI employs two families of magnets. Octupole magnets distributed around the ring introduce an amplitude dependent tune spread into the proton beam $dQ_x \propto x^2$. Particles with large betatron amplitudes have tunes closer to half-integer resonance of $Q_x = 26.5$ than those of small amplitude. Consequently, the phase space splits into stable and unstable areas, thereby providing a means for manipulating the extraction rate through control of the stable region in the phase space. Slow extraction from the MI proceeds as follows. The horizontal tune is raised towards the half-integer from its unperturbed value of $Q_x = 26.425$ to 26.485 using a family of quadrupole magnets. The strength of the octupoles is chosen such that the stable phase space area equals to the rms emittance of the circulating beam $\varepsilon_N \approx 0.55\pi$ mm mrad. The proton beam is just marginally stable staying just $\Delta Q_x = 0.015$ from the resonance. Finally, in the standard procedure, quadrupole magnets shift the tune onto the resonance and thus, the slow extraction takes place. If an electron lens is used to make the final tune shift of $\Delta Q_{EL} = 0.015$, then its current can be easily modulated in such a manner that the resonant extraction would occur only from particular batch(es) of proton bunches. In various operational modes, the Main Injector accelerates 4–6 batches of 84 proton bunches from 8 to 120–150 GeV with the cycle period of 1.5–4 s. The bunch spacing is 19 ns, so the duration of one batch is

Table 5.3 Main parameters of the electron lens for selective slow extraction of 120 GeV protons out of Fermilab's Main Injector [51]

Parameter	Symbol	Value	Unit
Length	L_{EL}	3.5	m
Current (max)	J_e	4.4	A
e-beam energy (max)	U_e	10	kV
Pulse repetition rate	f_0	88	kHz
Electron pulse rise time	t_e	~0.5	µs
Magnetic field in main solenoid	B_m	0.7	T
Magnetic field in gun solenoid	B_g	0.3	T
e-beam radius in main solenoid	a_e	3.3	mm
Cathode radius	a_c	5	mm
Tune shift by e-lens (max)	ΔQ_{EL}	0.015–0.25	

about 1.6 µs. Batches are separated by 532 ns gaps (28 bunches). Therefore, the electron current in the required electron lens system should have the following waveform: less than 532 ns risetime, at least 1.6 µs of the flat top (maximum current) and fall time shorter than 532 ns. The electrons must act on the chosen batch at every turn (revolution frequency $f_o = 88$ kHz) for the few seconds required for the slow extractions. The required peak electron current is about $J_e = 4.4$ A for a 10 keV electron beam energy and the interaction length of $L_{EC} = 3.5$ m. As we have shown in Sect. 2, these requirements are relatively straightforward and achievable with already developed and tested technology of the electron lenses. Main parameters of the electron lens for selective slow extraction out of the Main Injector are given in Table 5.3 from [51].

5.3   Beam-Beam Compensation in $e + e-$ Colliders with Electron Lenses

Beam-beam effects are setting severe limits on the luminosity of electron-positron colliders [52–54]. Given strong radiation damping the limiting beam-beam parameters for such colliders are usually in the range of $\xi_{e+e-} \sim 0.05$–0.1 [55, 56]—which are much higher than in hadron colliders—with record values reaching ~0.25 [44]. Besides the fast damping due to the synchrotron radiation, the most notable differences between the beam-beam phenomena for $e + e-$ and p-p collisions is that colliding beams are usually flat in the electron-positron colliders, beta functions at the IPs are not equal $\beta_y \ll \beta_x$, and beam energies and intensities are not always similar for electron and positrons, though in the most recently built and operated asymmetric B-factories all the beam-beam parameters are approximately the same $\xi_{y,e+} \approx \xi_{x,e+} \approx \xi_{y,e-} \approx \xi_{x,e-}$ [57].

As proposed and discussed in [58], the electron lenses can potentially be used for the beam-beam compensation in the $e + e-$ colliders. Indeed, for a non-linear (head-on) beam-beam compensation of the electron beams disrupted by collisions with positively charged positrons, one has to create a low energy electron beam which would generate the tune shift distribution similar to the *positron-induced electron tuneshift* distribution. In the case of matched beam-beam parameters like in the *B*-factories $\xi_{y,e-} \approx \xi_{x,e-}$, the Gaussian electron gun technology, as for the non-linear beam-beam compensation for the proton colliders, can be employed (see Sect. 3.2). If the symmetry condition does not hold $\xi_{y,e-} \neq \xi_{x,e-}$ then either round Gaussian electron beam with $j_e(r) \sim exp(-r^2/2\sigma^2)$ can be placed in a location with unequal beta-functions at the vertical and horizontal planes $\beta_{y,e-}/\xi_{y,e-} = \beta_{x,e-}/\xi_{x,e-}$ or (more cumbersome) a flat and Gaussian low energy electron beam can be generated out of a flat cathode. The issue with the latter option is that along the lens' interaction region, the flat electron beam will rotate due to its' own space charge forces, but that can be suppressed by having very strong magnetic field along the interaction region. (Definitely, the solenoid field has to be compensated to avoid transverse $x-y$ coupling in the high-energy electron beam.) As for the compensation in the positron beam disrupted by collisions with negatively charged high energy electrons, the shape of the electron lens beam needs to be reversed $j_e(r) \sim 1-exp(-r^2/2\sigma^2)$. Generation of such a beam might pose some challenges but this seemingly can be done the same way as for a hollow electron beam collimation— see Sect. 4.1. It was estimated in [58], that $L_e = 2$ m long electron lenses for head-on beam-beam compensation in the PEP-II B-factory [59] will require only about ~1 A of 10 kV electron current. It has to be noted that besides the electron lenses, several other methods to improve the beam-beam dynamics and increase the beam-beam parameters $\xi_{e+,e-}$ and therefore, the luminosity of $e+e-$ circular colliders have been proposed and some of them tested, including so called "round beams" [43, 44], "crab crossing" [75, 57] and "crab waist" [60–62].

5.4   Electron Beams to Control Transverse Instabilities

Collective transverse beam instabilities are of serious concern for operation of hadron supercolliders and set severe limits on the proton beam intensities. For example, the resistive wall impedance of the Lambertson injection magnets gave rise to the proton beam instabilities in the Tevatron [63] while the impedance of the particle collimators was one of the most stringent limitations on the LHC operation during the Run I [64, 65]. There are several ways to suppress these instabilities and one of the most effective is the Landau damping which requires introduction of spread of the betatron frequencies $\delta Q_{x,y}$, usually of the order of few 0.001 [66]. In practice, the Landau damping is usually implemented via employment of the octupole magnets which generate the tune spread proportional to square of the particle's amplitude $\delta Q_x \sim x^2$. The method has some limitations, though. E.g., the strength of the existing octupole circuits in the LHC is not always sufficient to keep the beam stable above certain proton bunch intensities. Moreover, even at their maximum strength, the octupoles significantly reduce the dynamic aperture of the

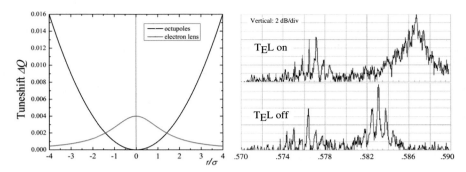

Fig. 5.6 (**a**)—*left*: illustration of the beam tune spreads induced by octupole magnets and by electron lens; (**b**)—*right*: the Tevatron beam Schottky monitor spectra showing spread of the betatron frequencies without (*bottom*) and with (*top*) the TEL-1 TEL-1 beam in the Tevatron (980 GeV protons, tune shift $\Delta Q_{EL} \sim 0.004$, additional tune spread $\delta Q_{EL} \sim 0.003$)—from [67]

Table 5.4 Electron beam requirements to generate a tune spread $\delta Q_{EL} \approx 0.004$ and stabilize $N_p = 2.3 \; 10^{11}$ protons/bunch in the LHC

Parameter	Symbol	Value	Unit
Length	L_{EL}	2.0	m
Number of electron lenses	N_{EL}	2	one/beam
Beta-functions at the e-lens	$\beta_{x,y}$	~200	m
Current (DC, max)	J_e	0.5–1	A
Electron beam profile		Gaussian	
e-beam radius in main solenoid	σ_e	0.3	mm
Magnetic field in main solenoid	B_m	6.5	T
Magnetic field in gun solenoid	B_g	0.2	T
Tune spread by e-lens (max)	δQ_{EL}	~0.004	

machine and the beam lifetime. The reason for that, as shown schematically in Fig. 5.6a, is that introduction of the required tune spread by the octupoles in the beam core results by default in significant non-linear fields (tuneshifts, etc.) for the particles at larger amplitudes. To avoid such lifetime degradation, it was proposed in [68] to employ an electron lens which would induce the tune spread solely in the core without degrading the dynamics of the particles in the transverse beam halo— see Fig. 5.6a. Table 5.4 lists the parameters of the electron lens to generate a tune spread $\delta Q_{EL} \approx 0.004$ and thus, stabilizing about 2.3 10^{11} protons/bunch. i.e., twice the design proton bunch intensity. Given the flexibility of the electron lenses they can be effectively used for the proton beam stabilization at all stages of collider operation—at injection, on the energy ramps, during the low-beta squeeze and, if necessary, in collisions.

The phenomenon of the increased tunespread by electron lens has been experimentally observed in the Tevatron. Figure 5.6b from [67] demonstrates that action of the misaligned TEL-1 electron beam resulted in significant widening of the synchro-betatron lines of the proton Schottky spectra by about $\delta Q_{EL} \sim 0.003$.

5.5 Beam-Beam Kicker

The aspiration of having ever higher beam currents and luminosities in limited footprint accelerator facilities naturally leads to development of strorage rings and colliders with operate with up to very large number of bunches (hundreds to thousands). The examples include *B*-factories [57, 59], *Tau-Charm* factories, damping rings for linear colliders, energy recovery linacs (ERLs), etc. The growing number of bunches usually results in reduced bunch-to-bunch distances, and could produce difficulties with multi-turn ejection/injection. Powerful multi-turn bunch-by-bunch ejection and injection kickers are needed for example, for the International Linear Collider (ILC) damping rings [69]. Conventional kickers with rise and fall times on the order of tens of ns cannot handle bunches separately if the bunch-to-bunch spacing is only a few meters. A novel method of a very fast kicker based on beam-beam forces was suggested in [70]. The method assumes impact of a high pulse current; low energy beam on bunches circulating in a storage ring. Ultimately, the method could allow for handling of separate bunches spaced by only a few tens of centimeters.

Figure 5.7 illustrates the principle of the beam-beam kicker operation. A low energy, high peak current, short-pulse electron beam is injected into the vacuum chamber of the storage ring. It moves over distance L_e along the orbit of the high energy beam (HEB) circulating in the ring. Both the electric and magnetic forces of the low energy electron beam kick the HEB while the two bunches pass each other. The electron beam then leaves the vacuum chamber to the beam dump before the arrival of the next high energy bunch. Therefore, the beam-beam kick duration is the larger of the electron pulse length t_e and electron propagation time $t_k \sim L_e/\beta_e c$ where c is the speed of light. The technology of the electron guns developed for the electron lenses—see Sect. 2—allows a minimal anode-cathode HV pulse of about 5–10 ns (limited by the time needed for electrons to travel from the cathode to the anode). When much shorter time scales are needed, either gridded thermionic guns

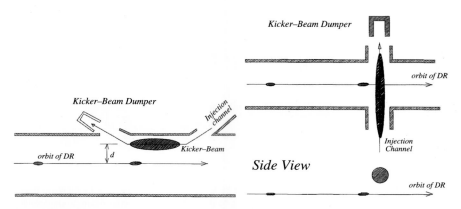

Fig. 5.7 Beam-beam kicker: (*left*) "head-on" configuration; (*right*) "cross" scheme [70]

or RF photo injectors can be employed. It also helps to operate with semi-relativistic beams $\beta_e \approx 1$ to attain kicker times of about a few ns in one-meter long systems. Numerical estimates [70] suggest that kick strengths approaching ~100 $G \cdot m$ are possible for such "head-on beam-beam kicker" (Fig. 5.7a) that made it a viable option for the TESLA $e + e-$ collider [71] damping ring injection and extraction. In the case where even shorter kicker-pulse durations are needed, the "cross" scheme (see Fig. 5.7b) could offer a way to sub-ns deflection times at the expense of reduced kicker strength $<10 G \cdot m$.

Recently, the design of the Medium Energy Electron Ion Collider (MEIC) [72] called for development of ultrafast beam-beam kicker for its ERL circulator cooler design which needs kicking electron bunches in and out of the circulator ring. The main parameters for such kicker are high repetition rate of 5–15 MHz, about 1.3 ns rise/fall time of the kick and 5 kV of integrated kicking voltage (~0.16 $G \cdot m$)—see a full list of parameters in Table 5.5.

Flat kicker beam can be produced out of a grid-operated DC (thermionic) electron gun with a round magnetized cathode followed by a "round-to-flat" beam transformation optics [73, 74]. The high energy (electron) beam will move at the speed of light across the non-relativistic kicking flat electron beam—see Fig. 5.8—at a very close distance to it. The resulting instant angular kick will be determined by integration of the transverse force over that passing time:

Parameter	Value	Unit
Circulating beam energy	33	MeV
Electron kicker beam energy	~0.3	MeV
Kicker repetition rate	5–15	MHz
Kicking angle	0.2	mrad
Kicking bunch length	15–50	cm
Kicking bunch width	0.5	cm
Kicking bunch charge	2	nC

Table 5.5 Design parameters of a beam-beam kicker for the Medium Energy Electron Ion Collider (MEIC) [72]

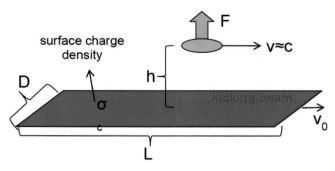

Fig. 5.8 Schematic drawing of beam-beam fast kicker (D is width of the kicking electron beam, L is its length, and σ is the surface electron charge density). Main high energy beam propagates at a distance h from the low energy kicking beam, which moves in the same direction to allow shorter kick duration [72]

$$\Delta\theta = \frac{2\pi r_e N_e}{\gamma \sigma_x}.$$ (5.6)

where N_e and σ_x are number of electrons and horizontal rms size of the kicker bunch and r_e is the electron classical radius. For the parameters listed in Table 5.5, the kick can reach the value of $\Delta\theta \approx 0.2$ mrad.

References

1. L.J. Laslett, BNL Report No. 7535 (1963) p. 154
2. C.E. Nielsen, A.M. Sessler, Rev. Sci. Instrum. **30**, 80 (1959)
3. B.W. Montague, CERN Report No. 68 (1968)
4. F. Sacherer, LNBL Report No. UCRL-18454 (1968)
5. I. Hofmann et al., Part. Accel. **13**, 145 (1983)
6. J. Struckmeier, M. Reiser, Part. Accel. **14**, 227 (1984)
7. S. Machida, Nucl. Instrum. Meth. Phys. Res. Sect. A **309**, 43 (1991)
8. S. Machida, Nucl. Instrum. Meth. Phys. Res. Sect. A **384**, 316 (1997)
9. I. Hofmann, Phys. Rev. E **57**, 4713 (1998)
10. S.Y. Lee, H. Okamoto, Phys. Rev. Lett. **80**, 5133 (1998)
11. A.V. Fedotov, J. Holmes, R.L. Gluckstern, Phys. Rev. ST Accel. Beams **4**, 084202 (2001)
12. J.A. Holmes et al., Phys. Rev. ST Accel. Beams **2**, 114202 (1999)
13. S.G. Anderson et al., Phys. Rev. ST Accel. Beams **5**, 014201 (2002)
14. A. Burov, Phys. Rev. ST Accel. Beams **12**, 044202 (2009)
15. A. Burov, V. Lebedev, Phys. Rev. ST Accel. Beams **12**, 034201 (2009)
16. M. Reiser, *Theory and Design of Charged Particle Beams* (Wiley, New York, 2008)
17. B. Zotter, in *Handbook of Accelerator Physics and Engineering*, ed. by A. Chao et al., (2nd Edition, World Scientific, Singapore, 2013), pp. 137–140
18. K. Hirata, J. Jowett (Eds.), *ICFA Beam Dynamics Newsletter* **20** (1999)
19. M. Aiba et al., in *Proc. 2007 IEEE PAC* (Albuquerque, NM, USA, 2007), p. 3390
20. *eRHIC Design Study: An Electron-Ion Collider at BNL*, http://arxiv.org/abs/1409.1633
21. Y. Zhang, J. Bisognano (Eds.), *Science Requirements and Conceptual Design for a Polarized Medium Energy Electron-Ion Collider at Jefferson Lab*, http://arxiv.org/abs/1209.0757
22. V. Shiltsev et al., in *Proceedings of IEEE NA-PAC'13* (Pasadena, CA, USA, 2013), p. 99
23. V. Shiltsev et al., in *Proceedings of IPAC'15* (Richmond, 2015), p. 4019
24. V. Shiltsev, in *Handbook of Accelerator Physics and Engineering*, ed. by A. Chao et al., (2nd Edition, World Scientific, Singapore, 2013), pp. 394–395
25. P.J. Bryant et al., Preprint CERN ISR-MA/75-54 (1975)
26. B.G. Logan et al., Nucl. Fusion **45**, 131 (2005)
27. G. Dimov, V. Chupriyanov, Part. Acc. **14**, 155 (1984)
28. A. Burov, G. Foster, V. Shiltsev, FNAL-TM-2125 (2000)
29. V. Litvinenko, G. Wang, Phys. Rev. ST Accel. Beams **17**(11), 114401 (2014)
30. Y. Alexahin et al., in *Proceedings of 2007 IEEE PAC* (Albuquerque, NM, USA, 2007), p. 3474
31. Y. Alexahin, V. Kapin, Fermilab document *beams-doc-3106* (2008), at http://beamsdoc.fnal.gov
32. V. Shiltsev et al., AIP Conf. Proc. **1086**, 649 (2009)
33. S. Machida, KEK note "Simulation results of space charge compensation with electron beams" (unpublished, 2001)
34. S. Nagaitsev et al., in *Proceedings of IPAC'12* (New Orleans, LA, USA, 2012), p. 16
35. A. Valishev et al., in *Proceedings of IPAC'12* (New Orleans, LA, USA, 2012), p. 1371

36. A. Valishev, S. Nagaitsev, V. Shiltsev, in *Proceedings of IPAC'15* (Richmond, VA, USA, 2015), MOPMA021
37. G. Stancari et al., in *Proceedings of IPAC'15* (Richmond, VA, USA, 2015), p. 46
38. E. Prebys et al., in *Proceedings of IPAC'15* (Richmond, VA, USA, 2015), p. 2627
39. V.V. Danilov, V.D. Shiltsev, Preprint FERMILAB-FN-0671 (1998)
40. V. Danilov, S. Nagaitsev, Phys. Rev. ST Accel. Beams **13**, 084002 (2010)
41. E.M. McMillan, University of California Report UCRL-17795 (1967)
42. E.M. McMillan, in *Topics in Modern Physics*, ed. by W.E. Brittin, H. Odabasi, (Colorado Associated University Press, Boulder, 1971), p. 219
43. V. Danilov, E. Perevedentsev, in *Proceedings of 1997 IEEE PAC* (Vancouver, Canada, 1997), p. 1759
44. D. Swartz et al., in *Proceedings of ICFA Mini-Workshop on Beam-Beam Effects in Hadron Colliders (BB2013, CERN, Geneva, Switzerland, 2013)*, ed. by W. Herr, G. Papotti, Preprint CERN-2014-004 (2014), p. 43
45. V. Shiltsev, in *Proceedings of IEEE 2007 PAC* (Albuquerque, NM, USA, 2007), p. 1159
46. V. Shiltsev, M. Chung, arXiv:1502.01736
47. G.H. Rees, P.J. Bryant, in *Handbook of Accelerator Physics and Engineering*, ed. by A. Chao et al., (2nd Edition, World Scientific, Singapore, 2013), pp. 382–387
48. See in M. Minty, F. Zimmermann, *Measurement and Control of Charged Particle Beams* (Springer, New York, 2003), pp. 230–238
49. S. Holmes, R. Gerig, D. Johnson, Part. Accel. **26**, 193 (1990)
50. C. Moore et al., in *Proceedings of 2001 IEEE PAC* (Chicago, IL, USA, 2001), p. 1559
51. V. Shiltsev, J. Marriner, in *Proceedings of 2001 IEEE PAC* (Chicago, IL, USA, 2001), p. 1468
52. K. Hirata, in *Handbook of Accelerator Physics and Engineering*, ed. by A. Chao et al. (2nd Edition, World Scientific, Singapore, 2013), pp. 169–174
53. K. Ohmi, Phys. Rev. E **62**(5), 7287 (2000)
54. R. Talman, Phys. Rev. ST Accel. Beams **5**, 081001 (2002)
55. J. Seeman, in *Nonlinear Dynamics Aspects of Particle Accelerators*, Lecture Notes in Physics Vol. **247** (Springer-Verlag, New York, 1985), p. 121
56. I. Koop, G. Tumaikin (Eds.), in *Proceedings of 3rd Advanced ICFA Beam Dynamics Workshop on Beam-Beam Effects in Circular Colliders* (Budker INP, Novosibirsk, Russia, 1989)
57. K. Oide, in *Elementary Particles—Accelerators and Colliders Series: Landolt-Börnstein: Numerical Data and Functional Relationships in Science and Technology—New Series*, ed. by H. Schopper, S. Myers, Sub volume 21C (Springer, New York, 2013), pp. 10.31–10.40
58. V. Shiltsev, in *Proceedings of 23rd Advanced ICFA Beam Dynamics Workshop on High Luminosity e + e− Colliders FACTORIES'2001* (Cornell University, Ithaca, NY, 2001); see, e.g., at http://www.lepp.cornell.edu/public/icfa/proceedings/index.html
59. *PEP-II Conceptual Design Report*, Preprint LBL-PUB- 5379, SLAC-418, CALT-68-1869, UCRL-ID-114055, UC- IIRPA-93-01 (1993)
60. P. Raimondi, in *Proceedings of the 2nd Workshop on Super B-Factory* (Frascati, Italy, 2006)
61. P. Raimondi, D. Shatilov, M. Zobov, INFN Report No. LNF-07/003; arXiv:physics/0702033
62. M. Zobov et al., Phys. Rev. Lett. **104**, 174801 (2010)
63. see Chapter 5 in V. Lebedev, V. Shiltsev (Eds.), *Accelerator Physics at the Tevatron Collider* (Springer, New York, 2014)
64. E. Métral et al., *Present Understanding of the Instabilities Observed at the LHC During Run I and Implications for HL-LHC*. in *Proceedings of 3rd Joint HiLumi LHC-LARP Annual Meeting* (Daresbury, UK, November 11–15, 2013)
65. E. Métral, *Initial Estimate of Machine Impedance*. No. CERN-ACC-2014-0005 (2014)
66. K.Y. Ng, *Physics of Intensity Dependent Beam Instabilities* (World Scientific, Singapore, 2006)
67. K. Bishofberger, *Successful Beam-Beam Tuneshift Compensation* (Ph.D. Thesis, UCLA, 2005)

68. V. Shiltsev, in *Proceedings of CARE-HHH-APD LHC-LUMI-06 Workshop "Towards a Roadmap for the Upgrade of the CERN & GSI Accelerator Complex"* (16–20 October 2006, Valencia, Spain), Yellow Report CERN-2007-002 (2007), p. 92
69. T.Behnke et al. (eds.), *The International Linear Collider Technical Design Report*, ILC-REPORT-2013-040 (2013)
70. V. Shiltsev, Nucl. Instrum. Meth. A **374**(2), 137–143 (1996)
71. F. Richard, J.R. Schneider, D. Trines, A. Wagner (eds.), *TESLA Technical Design Report*, Preprint DESY-2001-011 (2001)
72. Y. Zhang, J. Bisognano (eds.), *Science Requirements and Conceptual Design for a Polarized Medium Energy Electron-Ion Collider at Jefferson Lab, arxiv:1209.0757* (2012)
73. R. Brinkmann, Ya. Derbenev, K. Floettmann, TESLA Note 99-09 (1999)
74. P. Piot, Y.-E Sun, K.-J. Kim, Phys. Rev. ST Accel. Beams **9**, 031001 (2006)
75. T. Ieiri et al., Phys. Rev. ST Accel. Beams **12**, 064401 (2009)

Appendix: Theses on Physics, Technology and Applications of Electron Lenses

1. Christina Dimopoulou (2002), University of Lausanne/CERN
   "Design of a High-Perveance Electron Gun for Electron Cooling in the Low Energy Ion Ring (LEIR) at CERN and Non-Interceptive Proton Beam Profile Monitors using Ion or Atomic Probe Beams" (Ph.D. Thesis)
2. Kip Bishofberger (2005), UCLA/Fermilab
   "Successful Tevatron Beam-Beam Tuneshift Compensation" (Ph.D. Thesis)
3. Ulrich Dorda (2008), Technical University of Vienna/CERN
   "Compensation of long-range beam-beam interaction at the CERN LHC" (Ph.D. Thesis)
4. Ivan Morozov (2013), Novosibirsk State University, Russia
   "Modeling of Non-linear Elements in Storage Rings" (Master Thesis, in Russian)
5. Haroon Rafique, University of Huddersfield, UK (expected, 2016)
   "Hollow Electron Lenses as LHC Beam Halo Reducers" (Ph.D. Thesis)

© Springer Science+Business Media New York 2016
V.D. Shiltsev, *Electron Lenses for Super-Colliders*, Particle Acceleration and Detection, DOI 10.1007/978-1-4939-3317-4

Index

A

Abort gap, 19, 48, 73, 79, 98, 104, 156–160
Action variable, 114
Alignment, 64, 90, 91, 97, 105, 113, 123, 125, 126, 142, 143
Angle variable, 117
Antiproton(s), 1–3, 7, 9–14, 16, 19, 20, 26, 28, 30, 33, 35–49, 53, 57, 58, 72, 74, 77–80, 85–91, 93, 95, 96, 99, 103–105, 110, 113, 114, 116, 117, 120–124, 126, 127, 139, 142–147, 155–158, 160, 173

B

Batch, 72, 173, 174
Beam–beam effects
  bunch-by-bunch variations, 104
  coherent, 113
  head-on, 123
  incoherent, 167
  modeling, 117
  parasitic, 12, 14
  simulations, 88, 168
Beam–beam compensation
  head-on, 19, 20, 25, 31, 47, 114–134, 175
  long-range, 19, 21, 25, 26, 28, 33, 47, 61, 76, 85–113
  wire, 21
Beam-beam kicker, 21, 177–179
Beam-beam parameter, 10, 11, 85, 110, 114, 125, 126, 130, 133, 174, 175

Beam cooling
  electron, 16
  stochastic, 125, 126, 142
  synchrotron radiation, 13
Beam current
  average, 4
  peak, 156
Beam energy, 4, 6–8, 16, 27, 30, 49, 130, 148, 169, 170, 174, 178
Beam halo, 137, 151, 152, 163, 164, 169, 176
Beam lifetime, 99, 101, 104, 108, 116, 132, 133, 166, 175
  dynamic aperture, 117
Beam orbit, 7, 13, 44, 80, 85, 97, 129, 141, 146, 147, 156
Beam position monitor (BPMs), 53, 78–80, 103, 122
Beam separation, 11, 32, 88, 97, 124
Beta (betatron) function
  beta beating, 7
  collision point, 7
  phase advance, 115, 117
Betatron coupling, 34
Betatron phase, 115, 117, 119, 120, 130, 133, 143
Betatron tune
  diagnostics, 81
  Schottky monitor, 176
  tune scans, 123
BPMs. *See* Beam position monitors (BPMs)
Bunch lengthening, 14
Bunch oscillation, 41

© Springer Science+Business Media New York 2016
V.D. Shiltsev, *Electron Lenses for Super-Colliders*, Particle Acceleration and Detection, DOI 10.1007/978-1-4939-3317-4

Printed in the United States
By Bookmasters